ZEITSCHRIFT FÜR ANGEWANDTE MATHEMATIK UND MECHANIK

INGENIEURWISSENSCHAFTLICHE FORSCHUNGSARBEITEN

BAND 61 1981 HEFT 3

AUS DEM INHALT

HAUPTAUFSÄTZE

W. Werner: Penalty Function Methods for the Numerical Solution of Nonlinear Obstacle Problems with Finite Elements / **L. Jentsch:** Zur Lösung des linksseitigen Anfangswertproblems bei Systemen gewöhnlicher linearer Differentialgleichungen mit konstanten Koeffizienten (Existenz- und Eindeutigkeit, Approximation durch Lösungen singulär gestörter Systeme / **H. E. Salzer:** Rational Interpolation Using Incomplete Barycentric Forms / **R. Finn:** On the Laplace Formula and the Meniscus Height for a Capillary Surface / **R. Finn:** Addenda to My Paper "On the Laplace Formula and the Meniscus Height for a Capillary Surface" Preceding in this Issue / **H. Hilbig:** Existenzsatz für Potentialströmungen mit Totwassern um zwei geradlinige Hindernisse / **N. Phan-Thien:** On Stokes Flows in Channels and Pipes with Parallel Stationary Random Surface Roughness

KLEINE MITTEILUNGEN

BUCHBESPRECHUNGEN

THEORIA CUM PRAXI

AKADEMIE-VERLAG · BERLIN

ZAMM Bd. 61 Nr. 3 S. 133—208 Berlin, März 1981

EVP 18,— M 34115

INHALT

Hauptaufsätze

Seite

W. Werner: Penalty Function Methods for the Numerical Solution of Nonlinear Obstacle Problems with Finite Elements . . . 133

L. Jentsch: Zur Lösung des linksseitigen Anfangswertproblems bei Systemen gewöhnlicher linearer Differentialgleichungen mit konstanten Koeffizienten (Existenz- und Eindeutigkeit, Approximation durch Lösungen singulär gestörter Systeme . . . 141

H. E. Salzer: Rational Interpolation Using Incomplete Barycentric Forms . . . 161

R. Finn: On the Laplade Formula and the Meniscus Height for a Capillary Surface . . . 165

R. Finn: Addenda to My Paper "On the Laplace Formula and the Meniscus Height for a Capillary Surface" Preceding in this Issue . . . 175

H. Hilbig: Existenzsatz für Potentialströmungen mit Totwassern um zwei geradlinige Hindernisse . . . 179

N. Phan-Thien: On Stokes Flows in Channels and Pipes with Parallel Stationary Random Surface Roughness . . . 193

Kleine Mitteilungen

N. Anderson / A. M. Arthurs: Variational Solution of a Nonlinear Boundary Value Problem in the Theory of Power Law Fluids . . . 201

D. Karamanlidis: Über die numerische Berechnung ebener Rahmentragwerke im unter- und überkritischen Bereich nach der Finite-Element-Methode . . . 202

A. Granas / R. B. Guenther / J.W. Lee: A Note on the Thomas-Fermi Equation . . . 204

Buchbesprechungen . . . 205

Wir bitten, **Manuskriptsendungen** zweifach (Original und eine Kopie, sprachlich einwandfrei, Formeln mit Maschine oder in Druckschrift geschrieben) an folgende Anschrift zu richten:

Zeitschrift für Angewandte Mathematik und Mechanik
Institut für Mechanik der Akademie der Wissenschaften der DDR
DDR-1199 Berlin, Rudower Chaussee 5

Zu den Arbeiten, die als Hauptaufsätze bestimmt sind, ist auf gesondertem Blatt eine Zusammenfassung von 5 bis 10 Zeilen in englischer und (möglichst) deutscher und russischer Sprache beizufügen. Ausführliche Hinweise für die Autoren, um deren strikte Berücksichtigung gebeten wird, finden sich im Anschluß an das Inhaltsverzeichnis des Jahrganges **59** (1979).

Die Autoren erhalten von den Hauptaufsätzen 75, von den Kleinen Mitteilungen 25 Sonderdrucke ohne Berechnung, darüber hinaus weitere Sonderdrucke gegen Berechnung.

Der Verlag behält sich für alle Beiträge das Recht der Vervielfältigung, Vorbereitung und Übersetzung vor.

Bestellungen sind zu richten

- **in der DDR** an den Postzeitungsvertrieb unter Angabe der Kundennummer des Bestellenden oder an den AKADEMIE-VERLAG, DDR-1080 Berlin, Leipziger Str. 3—4
- **im sozialistischen Ausland** an eine Buchhandlung für fremdsprachige Literatur oder an den zuständigen Postzeitungsvertrieb
- **in der BRD und Berlin(West)** an eine Buchhandlung oder an die Auslieferungsstelle KUNST UND WISSEN, Erich Bieber OHG, D-7000 Stuttgart 1, Wilhelmstraße 4—6
- **in Österreich** an den Globus-Buchvertrieb, A-1201 Wien, Höchstädtplatz 3
- **in den übrigen westeuropäischen Ländern** an eine Buchhandlung oder an die Auslieferungsstelle KUNST UND WISSEN, Erich Bieber GmbH, CH-8008 Zürich/Schweiz, Dufourstraße 51
- **im übrigen Ausland** an den Internationalen Buch- und Zeitschriftenhandel; den Buchexport, Volkseigener Außenhandelsbetrieb der Deutschen Demokratischen Republik, DDR-7010 Leipzig, Postfach 160; oder an den AKADEMIE-VERLAG, DDR-1080 Berlin, Leipziger Straße 3—4

ZEITSCHRIFT FÜR ANGEWANDTE MATHEMATIK UND MECHANIK
Herausgeber und Chefredakteur: Prof. Dr. Günter Schmidt. Redaktion: Dr. Winfried Heinrich, Dr. Horst Weinert, Dipl.-Math. Friedhild Dudel, Helga Rühl, Institut für Mechanik der Akademie der Wissenschaften der DDR. Verlag: Akademie-Verlag, DDR-1080 Berlin, Leipziger Straße 3—4; Fernruf: 2236221 oder 2236229; Telex-Nr.: 114420; Bank: Staatsbank der DDR, Berlin, Kto.-Nr. 6836-26-20712. Anschrift der Redaktion: Institut für Mechanik der Akademie der Wissenschaften, DDR-1199 Berlin, Rudower Chaussee 5: Fernruf: 6802841/3639 u. 3643

Veröffentlicht unter der Lizenznummer 1282 des Presseamtes beim Vorsitzenden des Ministerrates der Deutschen Demokratischen Republik. Gesamtherstellung: VEB Druckerei „Thomas Müntzer", 5820 Bad Langensalza. Erscheinungsweise: Die Zeitschrift für Angewandte Mathematik und Mechanik erscheint monatlich. Die 12 Hefte eines Jahres einschließlich Tagungsheft bilden einen Band. Bezugspreis je Band 360,— M zuzüglich Versandspesen (Preis für die DDR 216,— M). Bezugspreis je Heft 30,— M (Preis für die DDR 18,— M). Bestellnummer dieses Heftes: 1009/61/3.

AN (EDV) 35937

ZAMM 61, 133 – 139 (1981)

W. WERNER

Penalty Function Methods for the Numerical Solution of Nonlinear Obstacle Problems with Finite Elements

Eine Klasse von Penalty-Methoden zur Lösung nichtlinearer Variationsungleichungen mit Hindernisnebenbedingungen

$\langle f - Au, v - u \rangle \leqq 0$ für alle $v \geqq \psi$

im Sobolev Raum $W^{1,p}(\Omega)$ wird untersucht. Die (nichtlinearen) Penalty-Gleichungen werden mit Hilfe der Finite Elemente Methode gelöst; die Konvergenzordnung dieses Verfahrens, welche von der Regularität der Lösung und den verwendeten Finiten Elementen abhängt, wird hergeleitet.

A class of penalty function methods for the solution of nonlinear variational inequalities with obstacles

$\langle f - Au, v - u \rangle \leqq 0$ for all $v \geqq \psi$

in the Sobolev space $W^{1,p}(\Omega)$ is studied. The (nonlinear) penalty equations are solved by finite element techniques; the order of convergence of this procedure which depends on the regularity of the solution as well as on the finite elements used is investigated.

Изучается класс методов штрафных функций для решения нелинейных вариационных неравенств с ограничениями в виде препятств

$\langle f - Au, v - u \rangle \leqq 0$ для всех $v \geqq \psi$

в пространстве Соболева $W^{1,p}(\Omega)$. Решаются (нелинейные) равенства с помощью метода финитных элементов. Выводится порядок сходимости способа зависящий от регулярности решения как и от применяемых финитных элементов.

1. Introduction

In many applications (see e.g. DUVAUT-LIONS [9]) problems of the following type arise:

$$\text{find} \quad u \in C \quad \text{such that} \quad \langle f - Au, v - u \rangle \leqq 0 \quad \text{for all} \quad v \in C . \tag{1.1}$$

Here C is a closed convex subset of a separable, reflexive BANACH space X, $A: X \to X'$ a monotone operator, $f \in X'$. In most cases (1.1) represents the variational formulation of a minimum problem in the BANACH space X, X e.g. a SOBOLEV space $W^{m,p}(\Omega)$.

In the sequel we restrict our interest to *obstacle problems* in the SOBOLEV space $W^{1,p}(\Omega)$ ($p \geqq 2$) with a T-monotone operator A. T-monotone operators have been introduced by BREZIS and STAMPACCHIA [5] and were further investigated by other authors (cf. MOSCO [20] and the references given there). T-monotony is intimately connected with the order structure of the SOBOLEV space $W^{1,p}(\Omega)$ which is an important tool for the characterization of the solution of obstacle problems (see HANOUZET and JOLY [14]).

Here we consider a penalty function method for the approximate solution of (1.1) which was used by LEWY and STAMPACCHIA [17] in existence and regularity investigations for linear problems; it was conjectured there that this penalty method "may be useful in a computational framework". The penalty equations are solved by finite element techniques here; we establish the rate of convergence of this procedure if suitable finite element spaces are used. In many special cases the order of convergence which is derived here corresponds to well known results of FALK [10], [11], MOSCO and STRANG [21], GLOWINSKI and MARROCO [13].

Our result indicates which order of convergence at least can be achieved by the application of finite element techniques to nonlinear obstacle problems; there are however cases where other methods for the solution of the obstacle problem yield a higher order (cf. BREZZI, HAGER, RAVIART [6]).

2. T-monotone operators in ordered Banach spaces

Let X be a real, separable, reflexive BANACH space and "$\geqq$" a partial order relation in X induced by a closed positive cone

$$P(X) := \{x \in X \mid x \geqq 0\} .$$

We assume that X is a lattice with respect to this ordering, that is any two vectors u and v in X have a common least upper bound denoted by $u \vee v$ and a common greatest lower bound denoted by $u \wedge v$ in X. Any vector $v \in X$ then can be decomposed as

$$v := v^+ - v^-$$

where $v^+ := v \vee 0$, $v^- := -(v \wedge 0)$. A subspace V of X is called a *sublattice* of the lattice X if for any two vectors $u, v \in V$ the elements $u \wedge v$ and $u \vee v$ belong to V. In the dual space V' of V one can define an order relation which is induced by the positive cone

$$P(V') := \{v' \in V' \mid \langle v', v \rangle \geqq 0 \quad \text{for all} \quad v \in P(V)\};$$

$\langle .,. \rangle$ being the *duality pairing* between V and V'.

2.1 Example: The SOBOLEV spaces $W^{1,p}(\Omega)$ with the partial ordering "$\leqq$" induced by the cone

$$P(W^{1,p}(\Omega)) := \{v \in W^{1,p}(\Omega) \mid v \geqq 0 \text{ a.e.}\}$$

is a lattice; $P(W^{1,p}(\Omega))$ is closed in $W^{1,p}(\Omega)$. The space $W_0^{1,p}(\Omega)$,

$$W_0^{1,p}(\Omega) := \text{"closure of } C_0^\infty(\Omega) \text{ in the SOBOLEV norm } \|v\|_{1,p} := \Big(\sum_{|\alpha| \leqq 1} \|D^\alpha v\|_p^p\Big)^{1/p}\text{"}$$

is a sublattice of $W^{1,p}(\Omega)$ (cf. HANOUZET and JOLY [14]).

2.2 Remark: The SOBOLEV spaces $W^{m,p}(\Omega)$, $m > 1$, with the partial ordering induced by the positive cone $P(L^p(\Omega))$ are no longer a lattice; essentially this is the reason why we must later on restrict our considerations on elliptic variational inequalities of *second* order.

2.3 Definition: Let X be a lattice, $V \subseteq X$ a sublattice; an operator $A: D(A) \subseteq X \to V'$ is called

(i) *monotone*, if for any two $u, v \in D(A)$: $\langle Au - Av, u - v\rangle \geqq 0$;

(ii) *strictly monotone*, if A is monotone and $\langle Au - Av, u - v\rangle = 0$ implies $u = v$;

(iii) *ϱ-monotone*, if there exists a continuous strictly increasing function ϱ: $[0, \infty) \to [0, \infty)$ with $\varrho(0) = 0$, $\lim_{x\to\infty} \varrho(x)/x = \infty$ such that for any $u, v \in D(A)$: $\langle Au - Av, u - v\rangle \geqq \varrho(\|u - v\|_X)$;

(iv) *strictly T-monotone*, if for any $u, v \in D(A)$: $(u - v)^+ \in V$ implies $\langle Au - Av, (u - v)^+\rangle \geqq 0$ and $\langle Au - Av, (u - v)^+\rangle = 0$ implies $v \geqq u$.

2.4 Remark: If $V \subseteq D(A)$ then any strictly T-monotone operator A is

(a) strictly monotone on V,

(b) of *monotone kind* on V: $Au \leqq Av$ implies $u \leqq v$.

2.5 Examples:

(a) Let $X = W^{1,p}(\Omega)$, $V = W_0^{1,p}(\Omega)$, $A: X \to V'$ be defined by

$$\langle Au, v\rangle := \langle A_0(\nabla u), \nabla v\rangle_{p,n} + \langle B_0 u, v\rangle_{p,1}$$

for all $u \in W^{1,p}(\Omega)$, $v \in W_0^{1,p}(\Omega)$ where $\langle.,.\rangle_{p,m}$ denotes the pairing between $(L^p(\Omega))^m$ and its topological dual $(L^q(\Omega))^m$ $(1/p + 1/q = 1)$;

$$A_0 : L^p(\Omega)^n \to L^q(\Omega)^n, \qquad B_0 : L^p(\Omega) \to L^q(\Omega)$$

are NEMYZKI-operators: for $w := (w_1, \dots, w_n) \in L^p(\Omega)^n$, $v \in L^p(\Omega)$, $x \in \bar\Omega$

$$(A_0 w)(x) := \big(a_1(x, w(x)), \dots, a_n(x, w(x))\big), \qquad (B_0 v)(x) := a_0\big(x, v(x)\big).$$

Here $a_i \in C(\bar\Omega \times \mathbb{R}^n)$ $(1 \leqq i \leqq n)$, $a_0 \in C(\bar\Omega \times \mathbb{R})$ satisfy the growth conditions

$$|a_i(x, y)| \leqq \alpha(x) + c \sum_{i=1}^n |y_i|^{p-1}, \qquad y \in \mathbb{R}^n, \qquad x \in \bar\Omega, \qquad 0 \leqq i \leqq n,$$

with $\alpha \in L^q(\Omega)$, $c > 0$. If

$$\sum_{i=1}^n \big(a_i(x, \xi) - a_i(x, \eta)\,(\xi_i - \eta_i)\big) > 0, \qquad a_0(x, y_0) \geqq a(x, y_1)$$

for any $x \in \bar\Omega$, $\xi \neq \eta \in \mathbb{R}^n$, $y_0 \geqq y_1$, *then A is strictly T-monotone.*

If furthermore

$$\sum_{i=1}^n (a_i(x, \xi) - a_i(x, \eta)\,(\xi_i - \eta_i) \geqq \gamma \sum_{i=1}^n |\xi_i - \eta_i|^p$$

for any $x \in \Omega$, $\xi \neq \eta \in \mathbb{R}^n$ with some $\gamma > 0$ then the restriction of A to $W_0^{1,p}(\Omega)$ is ϱ-monotone with respect to the function ϱ, $\varrho(x) := \gamma x^p$, if one provides $W_0^{1,p}(\Omega)$ with the norm

$$\|u\|_{1,p,0} := \left(\sum_{i=1}^n \left\|\frac{\partial}{\partial x_i} u\right\|_p^p\right)^{1/p}.$$

Throughout this paper Ω denotes a bounded simply connected open subset of $\mathbb{R}^n$ with LIPSCHITZ continuous boundary.

(b) A concrete instance of operators introduced in (a) are the so-called *pseudo-LAPLACians* A: $W^{1,p}(\Omega) \to W_0^{1,p}(\Omega)'$,

$$Au := -\sum_{i=1}^n \frac{\partial}{\partial x_i}\left(\left|\frac{\partial}{\partial x_i} u\right|^{p-2} \frac{\partial}{\partial x_i} u\right)$$

which are strictly T-monotone and ϱ-monotone on $W_0^{1,p}(\Omega)$.

2.6 Remark: In the sequel we investigate the solution of nonlinear variational inequalities in the SOBOLEV space $W_0^{1,p}(\Omega)$ with operators A of the form given in (2.5) (a); we always assume that A is strictly T-monotone and ϱ-monotone with respect to the function ϱ, $\varrho(x) := \gamma x^p$.

3. A class of penalty function methods for the solution of nonlinear obstacle problems

Let $\psi \in W^{1,p}(\Omega) \cap C(\bar{\Omega})$, $C := \{u \in W_0^{1,p}(\Omega) \mid u \geqq \psi\}$, where the ordering "$\geqq$" is to be understood in the sense of 2.1. We consider the variational inequality

$$\left.\begin{array}{l}\text{given } f \in L^q(\Omega) \text{ find } u \in C \text{ such that} \\ \langle f - Au, v - u\rangle \leqq 0 \quad \text{for all} \quad v \in C\,.\end{array}\right\} \tag{3.1}$$

If $\psi^+ \in W_0^{1,p}(\Omega)$ then C is a nonempty, closed convex subset of $W_0^{1,p}(\Omega)$; results concerning the existence and uniqueness of solution of (3.1) can be found e.g. in AUSLENDER [1], JEGGLE [15]. The regularity of solutions has been investigated by BREZIS [3], BREZIS and KINDERLEHRER [4], STAMPACCHIA [22], LIONS and STAMPACCHIA [19] and others; characterizations of the solutions are given by HANOUZET and JOLY [14]. Let us mention that our results remain valid for two-sided obstacle problems, too. Henceforth we assume that (3.1) possesses a unique solution.

As a first step we replace the variational inequality (3.1) by a sequence of variational *equalities*

$$\left.\begin{array}{l}\text{given } f \in L^q(\Omega) \text{ find } u_h \in W_0^{1,p}(\Omega) \text{ such that} \\ \langle f - Au_h, v\rangle = \langle P_h(\psi)\, u_h, v\rangle \quad \text{for all} \quad v \in W_0^{1,p}(\Omega)\,,\end{array}\right\} \tag{3.2}$$

where $P_h(4)\colon L^p(\Omega) \to L^q(\Omega)$ is some operator which may be interpreted as a penalty function, $h \in (0, 1)$ a real parameter. Methods of this type have been applied by several authors for the proof of existence and regularity results of solutions (see LIONS [18], BREZIS and STAMPACCHIA [5], STAMPACCHIA [22]).

Concerning $P_h(\psi)$ we make the following

3.1 Assumptions: $P_h(\psi)\, L^p(\Omega) \to L^q(\Omega)$ is defined by

$$\langle P_h(\psi)\, u, v\rangle := \int_\Omega \chi_h(u - \psi)\, v \,\mathrm{d}x \quad \text{for all} \quad u, v \in L^p(\Omega)$$

with a function $\chi_h\colon \mathbb{R} \to \mathbb{R}$ which satisfies the conditions:

(a) χ_h is monotone increasing;

(b) $\chi_h(x) = 0$ for $x \geqq \pi_h$; here π_h is a decreasing sequence of positive numbers which converges to zero as h approaches 0;

(c) χ_h is uniform LIPSCHITZ-continuous: for all $x, y \in \mathbb{R}$:

$$|\chi_h(x) - \chi_h(y)| \leqq c\pi_h^{-1}|x - y|\,.$$

3.2 Remarks:

(a) The solvability of the variational equality (3.2) under the above assumptions is well known (see e.g. ZEIDLER [24]).

(b) It is easily seen that $P_h(\psi)$ is not only a monotone operator on $L^p(\Omega)$ but indeed is *cyclically monotone* on $L^p(\Omega)$: For any $u_i \in L^p(\Omega)$, $k \in \mathbb{N}$ $(0 \leqq i \leqq k)$ with $u_0 = u_k$ it is

$$\sum_{i=1}^{k} \langle P_h(\psi)\, u_i, u_i - u_{i-1}\rangle \geqq 0\,.$$

(cf. BREZIS [2], JEGGLE [15]). This property of the penalty function will play a crucial role in the proof of lemma 4.7.

The following basic result mainly is due to STAMPACCHIA [22]:

3.3 Proposition: *Let*

(1) $A\colon W^{1,p}(\Omega) \to W_0^{1,p}(\Omega)'$ *be a strictly T-monotone and ϱ-monotone operator of the type introduced in* 2.5;

(2) $A\psi - f \in L^\infty(\Omega)$;

(3) $c_0 := |\chi_h(0)| \geqq \|(A\psi - f)^+\|_\infty$, c_0 *independent of h.*

Then the following estimates are valid:

(a) $\|u_h - u\|_p \leqq \pi_h$;

(b) $\|u_h - u\|_{1,p,0} \leqq c\pi_h^{1/p}$

($c > 0$ *being a constant independent of h*).

The *proof* consists of several steps:

(i) First of all we show that the solutions u_h of (3.2) belong to C: Since $(\psi - u_h)^+ \in W_0^{1,p}(\Omega)$ we may write

$$\begin{aligned} 0 &\leqq \langle A\psi - Au_h, (\psi - u_h)^+\rangle = \langle A\psi - f + P_h(\psi)\, u_h, (\psi - u_h)^+\rangle \\ &\leqq \langle A\psi - f + P_h(\psi)\, \psi(\psi - u_h)^+\rangle && \text{(by the monotony of } P_h(\psi)) \\ &\leqq 0\,. && \text{(by assumption (3))}\,. \end{aligned}$$

The strict T-monotony of A then implies $(\psi - u_h)^+ = 0$, i.e. $u_h \geqq \psi$.

(ii) By the definition of A (cf. 2.5) one has

$$\begin{aligned} 0 &\leqq \langle Au_h - A(u + \pi_h), (u_h - u - \pi_h)^+\rangle \leqq \langle Au_h - Au, (u_h - u - \pi_h)^+\rangle \\ &= \langle Au_h - f, (u_h - u - \pi_h)^+\rangle + \langle f - Au, \big(u + (u_h - u - \pi_h)^+\big) - u\rangle \\ &\leqq \langle Au_h - f, (u_h - u - \pi_h)^+\rangle \qquad \big(\text{by (3.1)}\big) \\ &= \langle -P_h(\psi)\, u_h, (u_h - u - \pi_h)^+\rangle = \int_\Omega |\chi_h(u_h - \psi)|\, (u_h - u - \pi_h)^+ \,dx \\ &= 0 \qquad \big(\text{by 3.1 (b)}\big)\,. \end{aligned}$$

Hence using the strict T-monotony of A one concludes:

$$\pi_h \geqq u_h - u \qquad \text{a.e. in } \Omega\,.$$

Furthermore $u - (u - u_h)^+ \in C$, whence

$$\begin{aligned} 0 &\leqq \langle Au_h - Au, -(u - u_h)^+\rangle = \langle Au_h - f, -(u - u_h)^+\rangle + \langle f - Au, \big(u - (u - u_h)^+\big) - u\rangle \\ &\leqq \langle Au_h - f, -(u - u_h)^+\rangle \qquad \big(\text{by (3.1)}\big) \\ &= \langle P_h(\psi)\, u_h, (u - u_h)^+\rangle \\ &\leqq 0 \qquad \big(\text{by 3.1 (a), (b)}\big)\,, \end{aligned}$$

hence: $(u - u_h)^+ = 0$, i.e. $u_h \geqq u$ a.e. in Ω. Thus (a) is established.

(iii) Since $u_h \in C$ we have

$$\begin{aligned} \gamma\|u - u_h\|_{1,p,0}^p &\leqq \langle Au_h - Au, u_h - u\rangle = \langle Au_h - f, u_h - u\rangle + \langle f - Au, u_h - u\rangle \\ &\leqq \langle Au_h - f, u_h - u\rangle = \int_\Omega |\chi_h(u_h - \psi)|\, (u_h - u)\,dx \\ &\leqq c\pi_h\mu(\Omega) \qquad \big(\text{by assumption (3)}\big)\,. \end{aligned}$$

Here $\mu(\Omega)$ denotes the usual LEBESGUE measure of Ω.

In actual computations it may be necessary to replace the original obstacle by some approximation; usually ψ is interpolated. Therefore we need the following perturbation result:

3.4 Lemma: *Let $\psi_h \in W^{1,p}(\Omega) \cap C(\bar\Omega)$, $h \in (0, 1)$, $\|\psi_h - \psi\|_\infty \leqq c\pi_h$; let us denote by $u_h(\psi_h)$ the solution of the variational equation* (3.2) *with ψ replaced by ψ_h. Then*

(a) $\|u - u_h(\psi_h)\|_p \leqq c_1\pi_h$,

(b) $\|u - u_h(\psi_h)\|_{1,p,0} \leqq c_2\pi_h^{1/p}$

for some positive constants c_1, c_2 independent of h.

3.5 Remark: Note that ψ_h may violate the assumptions of proposition 3.3; it even might be

$$\{v \in W_0^{1,p}(\Omega) \mid v \geqq \psi_h\} = \emptyset$$

since we do not demand "$\psi_h \leqq 0$ on $\partial\Omega$".

Proof of 3.4: In the sequel "c" denotes a generic constant which may be different at each instance.

It obviously suffices to show that

(a)′ $\|u_h - u_h(\psi_h)\|_p \leqq c\|\psi_h - \psi\|_\infty$,

(b)′ $\|u_h - u_h(\psi_h)\|_{1,p,0} \leqq c\|\psi_h - \psi\|_\infty^{1/p}$

Let $k := \|(\psi_h - \psi)^+\|_\infty$; then

$$\begin{aligned} 0 &\leqq \langle Au_h(\psi_h) - A(u_h + k), \big(u_h(\psi_h) - u_h - k\big)^+\rangle \leqq \langle Au_h(\psi_h) - Au_h, \big(u_h(\psi_h) - u_h - k\big)^+\rangle \\ &= \int_\Omega \big(\chi_h(u_h - \psi) - \chi_h(u_h(\psi_h) - \psi_h)\big)\big(u_h(\psi_h) - u_h - k\big)^+ \,dx \\ &\leqq 0 \qquad (\text{by the monotony of } \chi_h)\,. \end{aligned}$$

Hence: $u_h(\psi_h) - u_h \leqq k := \|(\psi_h - \psi)^+\|_\infty$.

Analogously one shows: $u_h(\psi_h) - u_h \geqq -\|(\psi_h - \psi)^-\|_\infty$ a.e. in Ω; thus we have proved (a)′.

The proof of (b)′ is similar to the proof of 3.3 (b) and therefore is omitted.

4. Finite element approximations for the obstacle problem

For the approximate solution of 3.2 we use the GALERKIN method with subspaces of $W_0^{1,p}(\Omega)$ which constitute a so-called $(k, 1)$-system:

4.1 Definition: A family of finite dimensional subspaces $S_h \subseteq W_0^{1,p}(\Omega) \cap C(\bar\Omega)$ $\big(h \in (0, 1)\big)$ is called a $(k, 1)$-*system* in $W_0^{1,p}(\Omega)$, if for any $v \in W_0^{1,p}(\Omega) \cap W^{r,p}(\Omega)$ $(r > 1)$ there is a $V_h \in S_h$ with

$$\|v - V_h\|_{W^{s,p}(\Omega)} \leqq ch^{t-s}\|v\|_{W^{t,p}(\Omega)}$$

where $s \in \{0, 1\}$, $t := \min(k, r)$, c independent of v and h.

4.2 Remark: Most finite element spaces which are used in practice form a $(k, 1)$-system (see. e.g. CIARLET [8]).

With respect to the operator A we need one further assumption:

4.3 Assumption: A is locally LIPSCHITZ-continuous, i.e. for any $u, v \in W_0^{1,p}(\Omega)$ we have

$$||Au - Av||_{W_0^{1,p}(\Omega)'} \leqq cR^{p-2}||u - v||_{1,p,0}$$

where $R := \max(||u||_{1,p,0}, ||v||_{1,p,0})$, c independent of u and v.

4.5 Remark: Assumption 4.3 is fulfilled e.g. for the pseudo-LAPLACians (cf. 2.5); it obviously reduces to the standard assumption of LIPSCHITZ-continuity in the Hilbert space case $p = 2$.

The definition of S_h and lemma 3.4 yield

4.6 Lemma: *Let*

$$(1)\quad u \in W_0^{1,p}(\Omega) \cap W^{r,p}(\Omega) \quad (r > 1), \qquad (2)\quad \overline{\lim_{h\to 0}}\, \pi_h^{-1}||\psi_h - \psi||_\infty < \infty;$$

then there is a $V_h \in S_h$ *such that*

$$||V_h - u_h(\psi_h)||_p \leqq c(h^t + \pi_h), \qquad ||V_h - u_h(\psi_h)||_{1,p,0} \leqq c(h^{t-1} + \pi_h^{1/p})$$

with a positive constant c independent of h.

Now let $U_h \in S_h$ be the solution of the GALERKIN equations

$$\langle f - AU_h, V_h\rangle = \langle P_h(\psi_h)\, U_h, V_h\rangle \quad \text{for all} \quad V_h \in S_h. \tag{4.1}$$

Our aim is to choose the penalty parameter π_h in dependence of h such that the order of convergence of

$$||u - U_h||_{1,p,0}$$

is as large as possible. Under the assumptions of 4.6 we have

4.7 Lemma: *If* $\overline{\lim_{h\to 0}}\, h^{2t}\cdot \pi_h^{-1} < \infty$ *then the solutions* U_h *of* (4.1) *can be estimated in the following way:*

$$||U_h - u_h(\psi_h)||_{1,p,0} \leqq c(\varepsilon_1(h)^{1/(p-1)} + \varepsilon_2(h)^{1/p})$$

where we used the notation

$$\varepsilon_1(h) := h^{t-1} + \pi_h^{1/p}, \qquad \varepsilon_2(h) := h^t + \pi_h + \pi_h^{-1}\cdot h^{2t}.$$

Proof: In order to simplify notation we write

$$U := U_h, \qquad v := u_h(\psi_h), \qquad \chi := \chi_h.$$

From (4.1) we get

$$\langle f - AU, V_h - U\rangle - \int_\Omega \chi(U - \psi_h)\,(V_h - U)\,dx \quad \text{for all} \quad V_h \in S_h,$$

(3.2) (with obstacle ψ_h) yields

$$\langle f - Av, V_h - U\rangle = \int_\Omega \chi(v - \psi_h)\,(V_h - U)\,dx \quad \text{for all} \quad V_h \in S_h \subseteq W_0^{1,p}(\Omega).$$

Subtracting both equations one gets

$$\langle AU - Av, V_h - U\rangle = \int_\Omega \big(\chi(v - \psi_h) - \chi(U - \psi_h)\big)\,(V_h - U)\,dx.$$

Hence for any $V_h \in S_h$

$$\begin{aligned}\gamma||U - v||_{1,p,0}^p &\leqq \langle AU - Av, U - v\rangle = \langle AU - Av, V_h - v\rangle + \langle AU - Av, U - V_h\rangle\\ &= \langle AU - Av, V_h - v\rangle + \int_\Omega \big(\chi(U - \psi_h) - \chi(v - \psi_h)\big)\,(V_h - U)\,dx\\ &\leqq cR^{p-2}||U - v||_{1,p,0}\,||V_h - v||_{1,p,0} + \int_\Omega \big(\chi(U - \psi_h) - \chi(v - \psi_h)\big)\,(V_h - U)\,dx\end{aligned}$$

where $R := \max(||U||_{1,p,0}, ||v||_{1,p,0})$.

From the cyclical monotony of $P_h(\psi_h)$ (cf. 3.2) we conclude:

$$\int_\Omega \big(\chi(U - \psi_h) - \chi(v - \psi_h)\big)\,(V_h - U)\,dx \leqq \int_\Omega \big(\chi(V_h - \psi_h) - \chi(v - \psi_h)\big)\,(V_h - v)\,dx \quad \text{for all} \quad V_h \in S_h.$$

so that we finally get

$$||U - v||_{1,p,0}^p \leqq c(R^{p-2}||U - v||_{1,p,0}\,||V_h - v||_{1,p,0} + \pi_h^{-1}||V_h - v||_p^2) \quad \text{for all} \quad V_h \in S_h.$$

Without loss of generality we can assume that

$$||U_h||_{1,p,0} = ||U||_{1,p,0} \geqq c||v||_{1,p,0} = c||u_h(\psi_h)||_{1,p,0} \qquad (c > 0)$$

since by lemma 3.4 $u_h(\psi_h)$ is convergent for $h \to 0$.

Hence we have — returning now to the original notation:

$$||U_h - u_h(\psi_h)||_{1,p,0}^p \leqq c\big(||U_h||_{1,p,0}^{p-2}\,||U_h - u_h(\psi_h)||_{1,p,0}\,||V_h - u_h(\psi_h)||_{1,p,0} + \pi_h^{-1}||V_h - u_h(\psi_h)||_p^2\big)$$

$(V_h \in S_h$ arbitrary)

so that

$$||U_h - u_h(\psi_h)||^p_{1,p,0} \leq c(||U_h||^{p-2}_{1,p,0} ||U_h - u_h(\psi_h)||_{1,p,0}\, \varepsilon_1(h) + \varepsilon_2(h))$$

(by 4.6); the constant c being dependent on the data of the variational inequality. The last inequality is of the type

$$x^p \leq ax + b \qquad (a, b > 0);$$

the positive solutions of this inequality can be estimated by

$$0 \leq x \leq \max\left((2a)^{1/(p-1)}, (2b)^{1/p}\right).$$

Hence

$$||U_h - u_h(\psi_h)||_{1,p,0} \leq c(||U_h||^{(p-2)/(p-1)}_{1,p,0}\, \varepsilon_1(h)^{1/(p-1)} + \varepsilon_2(h)^{1/p}).$$

Especially we conclude that $\{||U_h||_{1,p,0} \mid h \in (0, \bar{h})\}$ is uniformly bounded for $\bar{h}$ sufficiently small; assume that $\{||U_h||_{1,p,0} \mid h \in (0, \bar{h})\}$ is unbounded: then $\lim_{h\to 0} ||U_h - u_h(\psi_h)||_{1,p,0}/||U_h||_{1,p,0} = 0$.

On the other hand

$$\begin{aligned} 1 &= \overline{\lim_{h\to 0}}\, (1 + ||u_h(\psi_h)||_{1,p,0}/||U_h - u_h(\psi_h)||_{1,p,0})^{-1} \\ &= \overline{\lim_{h\to 0}}\, ||U_h - u_h(\psi_h)||_{1,p,0}/(||U_h - u_h(\psi_h)||_{1,p,0} + ||u_h(\psi_h)||_{1,p,0}) \\ &\leq \lim_{h\to 0} ||U_h - u_h(\psi_h)||_{1,p,0}/||U_h||_{1,p,0} = 0 \end{aligned}$$

which is a contradiction. Thus the proposed inequality is established.

4.8 Corollary: *If in addition to the assumptions of the foregoing lemma* 4.7

$$\lim_{h\to 0} h^{2t} \cdot \pi_h^{-1} = 0\,,$$

then $\lim_{h\to 0} U_h = u$ *in* $W_0^{1,p}(\Omega)$, *where* u *is the unique solution of* (3.1).

We are now able to state our main result:

4.9 Proposition: *Let* $u \in W_0^{1,p}(\Omega) \cap W^{r,p}(\Omega)$ $(r > 1, p \geq 2)$ *be the unique solution of the variational inequality* (3.1), *let all the assumptions on* A *which have been made be fulfilled; let* S_h *be a* $(k, 1)$*-regular system. If*

$$||\psi_h - \psi||_\infty = O(h^t)\,, \qquad \pi_h = O(h^t)\,,$$

then the solutions of the penalty equations (3.2) *satisfy the asymptotic estimate:*

$$||U_h - u||_{1,p,0} = O(h^\varkappa)$$

where

$$\varkappa = \begin{cases} \dfrac{t-1}{p-1} & \text{if } t \leq q \\[2ex] \dfrac{t}{p(p-1)} & \text{if } t \geq q \end{cases}\,, \qquad 1/p + 1/q = 1\,, \qquad t = \min(k, r)\,.$$

Proof:

$$\begin{aligned} ||U_h - u||_{1,p,0} &\leq ||U_h - u_h(\psi_h)||_{1,p,0} + ||u_h(\psi_h) - u||_{1,p,0} \\ &\leq c(h^{t-1} + h^{t/p})^{1/(p-1)} \qquad \text{(by 3.4, 4.7)} \\ &= O(h^\varkappa)\,, \qquad \text{with } \varkappa \text{ as defined above}\,. \end{aligned}$$

4.10 Remarks:

(a) For linear elliptic variational inequalities of second order arising from a constraint optimization problem the order of convergence of GALERKIN methods using linear finite elements has been investigated by FALK [10], [11] and Mosco and STRANG [21]; in that case one has $p = 2$ and — under suitable regularity assumptions on the data — $r = 2$. These authors show $O(h)$-convergence of their finite element approximations.

(b) BREZZI, HAGER, RAVIART [6] established the $O(h^{1.5-\varepsilon})$-convergence in the case of quadratic finite elements; their work covers the case $p = 2$, $r = 2.5 - \varepsilon$, ε arbitrary small (the quantity 2.5-ε originates from a regularity result of BREZIS [3]). This order of convergence is not reproduced by 4.9 under corresponding assumptions. Proposition 4.9 however also covers more general variational inequalities, such as nonlinear obstacle problems, which do not fall into the range of the results cited above.

(c) The computational advantage of penalty-methods of the type investigated here is their simple applicability: one can apply any standard finite element routine for the solution of nonlinear boundary value problems to handle linear or nonlinear obstacle problems, too. This device is justified by our proof of convergence which at the same time gives concrete hints how to choose the penalty-function and the penalty-parameter.

References

1 AUSLENDER, A., Problèmes de Minimax via l'analyse convexe et les inégalités variationels, Lecture Notes in Economics 77, Springer, Berlin-Heidelberg-New York 1972.
2 BREZIS, H., Opérateurs maximaux monotones, North Holland, Amsterdam-London 1973.
3 BREZIS, H., Problèmes unilatéraux, J. Math. Pures et Appl. **51**, 1–168 (1972).
4 BREZIS, H.; KINDERLEHRER, D., The smoothness of solutions to nonlinear variational inequalities, Indiana Univ. Math. Journ. **23**, 831–844 (1974).
5 BREZIS, H.; STAMPACCHIA, G., Sur la régularité de la solution d'inéquations elliptiques, Bull. Soc. Math. France **96**, 153–180 (1968).
6 BREZZI, F.; HAGER, W. W.; RAVIART, P. A., Error estimates for the finite element solution of variational inequalities I, Numer. Math. **28**, 431–443 (1977).
7 BREZZI, F.; HAGER, W. W.; RAVIART, P. A., Error estimates for the finite element solution of variational inequalities II, Numer. Math. **31**, 1–16 (1978).
8 CIARLET, P. G., Numerical Analysis of the Finite Element Method, North-Holland, Amsterdam-London 1977.
9 DUVAUT, G.; LIONS, J. L., Les inéquations en mécanique et en physique, Dunod, Paris 1972.
10 FALK, R. S., Error estimates for the approximation of a class of variational inequalities, Math. Comput. **28**, 963–971 (1974).
11 FALK, R. S., Approximation of an elliptic boundary value problem with unilateral constraints, RAIRO **9**, 5–12 (1975).
12 GLOWINSKI, R.; LIONS, J. L.; TREMOLIERES, R., Analyse numérique des inéquations variationelles. Tome 1, 2, Dunod, Paris 1976.
13 GLOWINSKI, R.; MARROCO, A., Sur l'approximation, par éléments finis d'ordre un, et la résolution, par pénalisation-dualité, d'une classe de problèmes de DIRICHLET non-linéaire, RAIRO **9**, 41–76 (1975).
14 HANOUZET, B.; JOLY, J. L., Méthodes d'ordre dans l'interprétation de certaines inéquations variationelles et applications, J. Funct. Analysis **34**, 217–249 (1979).
15 JEGGLE, H., Nichtlineare Funktionalanalysis, Teubner, Stuttgart 1979.
16 KLUGE, R., Nichtlineare Variationsungleichungen und Extremalaufgaben, Deutscher Verlag der Wissenschaften, Berlin 1979.
17 LEWY, H.; STAMPACCHIA, G., On the regularity of the solution of a variational inequality, Comm. Pure Appl. Math. **22**, 153–188 (1969).
18 LIONS, J. L., Quelques méthodes de résolution des problèmes aux limites non-linéaires, Dunod, Paris 1969.
19 LIONS, J. L.; STAMPACCHIA, G., Variational inequalities, Commun. Pure Appl. Math. **20**, 493–519 (1967).
20 MOSCO, U., Implicit variational problems and quasivariational inequalities, in: GOSSEZ, J. P. (ed.): Nonlinear Operators and the Calculus of Variations, Lecture Notes in Mathematics 543, Springer, Berlin-Heidelberg-New York 1977.
21 MOSCO, U.; STRANG, G., One sided approximation and variational inequalities, Bull. Amer. Math. Soc. **80**, 308–312 (1974).
22 STAMPACCHIA, G., Variational inequalities, in: GHIZZETTI, A. (ed.): Theory and Applications of Monotone Operators, Proc. Nato Advanced Study Institute, 1969.
23 WERNER, W., Über ein Penalty-Verfahren zur Lösung nichtlinearer elliptischer Variationsungleichungen mit Hilfe der Finite Elemente Methode, Mainz 1978.
24 ZEIDLER, E., Vorlesungen über nichtlineare Funktionalanalysis I, II, III Teubner, Leipzig 1977.

Eingereicht: 18. 3. 1980

Anschrift: Dr. WILHELM WERNER, Fachbereich 17 der Johannes Gutenberg-Universität, D 6500 Mainz, BRD

ZAMM 61, 141–159 (1981)

L. JENTSCH

Zur Lösung des linksseitigen Anfangswertproblems bei Systemen gewöhnlicher linearer Differentialgleichungen mit konstanten Koeffizienten (Existenz und Eindeutigkeit, Approximation durch Lösungen singulär gestörter Systeme)

Es wird eine mathematisch exakte Formulierung des linksseitigen Anfangswertproblems in einer Klasse von Distributionen gegeben, die von regulären Distributionen mit positivem Träger und der δ-Distribution und ihren Ableitungen erzeugt wird. Zur Ermittlung von Lösungen ist die Methode der Laplace-Transformation geeignet. Die wesentliche Voraussetzung für die Existenz und die einzige Voraussetzung für die Eindeutigkeit der Lösung ist, daß das charakteristische Polynom nicht identisch Null ist. Danach werden Sätze über die Approximation der distributionentheoretischen Lösung des linksseitigen Anfangswertproblems, die für $t > 0$ als gewöhnliche Funktion interpretiert werden kann, durch Lösungen klassischer Anfangswertprobleme bewiesen. Dabei werden δ-Distributionen durch reguläre Distributionen und anormale Systeme durch singulär gestörte normale Systeme approximiert. Für die Konvergenz der Lösungen der singulär gestörten Probleme gegen die distributionentheoretische Lösung für $t > 0$ sind gewisse Stabilitätskriterien hinreichend.

A mathematical exact formulation of the left-side initial value problem in a class of distributions consisting of regular distributions with positive support and δ-distributions and their derivatives is given. The method of Laplace transform is properly to find out the solution. The essential assumption for the existence and the only assumption for the uniqueness of the solution is that the characteristic polynomial is not identically zero. After this theorems are proved concerning the approximation of the distributional solution of the left-side initial value problem, which for $t > 0$ may be interpreted as a usual function, by solutions of classical initial value problems. There δ-distributions are approximated by regular distributions and anormal systems by singular perturbed normal systems. The solutions of the singular perturbed problems converge to the distributional solution for $t > 0$, when some conditions of stability are fulfilled.

Даётся математически точная формулировка решения системы дифференциальных уравнений с левыми начальными условиями в некотором классе обобщённых функций, которые образуются регулярными распределениями с положительным носителем и δ-функцией и её производными. Для решения такой системы применяется преобразование Лапласа. Условие, что характеристический полином не тождественно равняется нулю, является существенным предположением существования и единственным предположением единственности решения. Далее, доказываются теоремы аппроксимации решения системы дифференциальных уравнений с левыми начальными условиями, найденного в области обобщённых функций, которые для $t > 0$ можно рассматривать как обычную функцию, решениями классических проблем. При этом δ-функции аппроксимируются регулярными функциями и ненормальные системы сингулярно возмущёнными системами. В случае $t > 0$ некоторые условия устойчивости являются достаточными для сходимости решения сингулярно возмущённой системы к решению, найденному в области обобщённых функций.

1. Einleitung

Systeme gewöhnlicher Differentialgleichungen treten z. B. bei elektrischen Netzwerken auf. Der Zustand ist für die Zeit $t < 0$ bekannt, und gesucht ist die Lösung für $t > 0$, bekannt sind also immer die linksseitigen Grenzwerte für $t = -0$. Man hat es demzufolge bei den Anwendungen ausnahmslos mit dem linksseitigen Anfangswertproblem zu tun. Ist das System *normal*, d. h. ist die Koeffizientendeterminante det $A^{(n)}$ bei den höchsten Ableitungen von Null verschieden, so stimmen bei vernünftiger rechter Seite die rechts- und linksseitigen Grenzwerte überein, und man kann das rechtsseitige Anfangswertproblem z. B. bequem mit Hilfe des Differentiationssatzes für die klassische LAPLACE-Transformation lösen. Bei den praktischen Anwendungen kommen jedoch auch *anormale* Systeme (det $A^{(n)} = 0$) vor (s. [1], Fußnote 4, [2], S. 102), bei denen die rechtsseitigen Grenzwerte nicht mit den linksseitigen Grenzwerten übereinzustimmen brauchen. Die aufgeworfene Problematik ist nicht neu (s. etwa [1], [2] S. 93–104, [3] S. 318–328, [4], [5], [6] S. 336–343, [7] S. 115–123, [8] S. 49–53, 67–69). In [2], [5], [9] werden solche Probleme mit der Operatorenrechnung behandelt. In [1] wird eine vollständige Theorie selbst im Fall, daß das charakteristische Polynom $\Delta(p) \equiv 0$ ist, gegeben. Ist $\Delta(p) \not\equiv 0$, so ist im Originalbereich für $t > 0$ eine eindeutige Lösung festgelegt, der die zurücktransformierbaren Anteile der Bildlösung entsprechen.

Ähnlich wie in [4] und in [6] vorgeschlagen fassen wir die Funktionen des Originalbereichs als reguläre Distributionen auf. Gegenüber [4], wo Lösungen nur im Bereich der Sprungfunktionen gesucht sind, erweitern wir diesen Bereich durch die δ-Distribution und ihre Ableitungen und wenden darauf LAPLACE-Transformation an. Damit erfassen wir auch, was bei $t = 0$ passiert, und das Vorgehen in [1] findet eine natürliche Rechtfertigung. Außerdem können wir damit die DIRACsche δ-Funktion und Ableitungen von ihr als Störfunktion einbeziehen.

Das Hauptanliegen der Arbeit jedoch sind Sätze, die die mit der Methode der verallgemeinerten LAPLACE-Transformation bzw. Operatorenrechnung gefundene Lösung für $t > 0$ als Grenzwert der Lösung eines klassischen Anfangswertproblems nachweisen, indem einerseits $\delta^{(j)}(t)$ durch klassische Funktionen $h_\eta^{(j)}(t)$ approximiert wird und andererseits ein anormales System als Grenzfall eines singulär gestörten normalen Systems aufgefaßt wird. An einem Beispiel wird eine solche Grenzbetrachtung in [2], S. 103 durchgeführt. In [10] wird für eine spezielle singuläre Störung bei einem System 1. Ordnung mit konstanten rechten Seiten ein Approximationssatz bewiesen, wenn das charakteristische Polynom des Ausgangs- und des gestörten Systems nur einfache Nullstellen hat. Auch in den von uns betrachteten allgemeineren Fällen ergibt sich als wesentliche Konvergenzbedingung, daß die Realteile der im charakteristischen Polynom des ε-gestörten Systems zusätzlich auftretenden Wurzeln für $\varepsilon \to +0$ gegen $(-\infty)$ gehen. Ist diese Bedingung erfüllt, sprechen wir von einem *stabil gestörten System.* In [11] ist gezeigt, daß jedes anormale System 1. Ordnung mit $\Delta(p) \not\equiv 0$ stets als Grenzfall eines stabil gestörten normalen Systems aufgefaßt werden kann. Zum Schluß sei noch vermerkt, daß die bei singulär gestörten Problemen auftretenden Grenzschichten beim Grenzübergang $\varepsilon \to +0$ auf $t = 0$ zusammenschrumpfen und sich in den in der Lösung des entarteten Problems vorkommenden δ-Distributionen wiederfinden.

2. Das linksseitige Anfangswertproblem

Es sei $\mathcal{D}$ der lineare Raum aller auf der reellen Achse R^1 unendlich oft differenzierbaren Funktionen $\varphi(x)$ mit kompaktem Träger und $\mathcal{D}' = \mathcal{D}_1'$ der lineare Raum aller auf $\mathcal{D}$ definierten reellen linearen stetigen Funktionale (Distributionen). Es bezeichne $D^k y$ die k-te Derivierte von $y \in \mathcal{D}'$ (zu den Begriffen aus der Distributionentheorie s. etwa [6], S. 451–469). Es sei $\mathcal{D}_m'$ der lineare Raum aller Vektoren $y = (y_1, \dots, y_m)^\mathsf{T}$ mit $y_j \in \mathcal{D}', j = 1, \dots, m$, und $D^k y = (D^k y_1, \dots, D^k y_m)^\mathsf{T}$. Die von einer lokal integrablen Funktion $f(t)$ erzeugte Distribution werde mit $[f]$, die δ-Distribution mit $[\delta]$ und die k-te Derivierte von $[\delta]$ mit $[\delta^{(k)}]$ bezeichnet. Für die Berechnung der k-ten Derivierten einer durch eine Sprungfunktion erzeugten Distribution gilt der

Satz 2.1 (s. etwa [6], S. 459): *Es besitze $f(t)$ für $t \neq 0$ Ableitungen bis zur k-ten Ordnung, $f^{(k)}(t)$ sei für $t \neq 0$ stetig und lokal integrabel, dann gilt*

$$D^k[f] = [f^{(k)}] + \sum_{j=0}^{k-1} \left(f^{(k-j-1)}(+0) - f^{(k-j-1)}(-0)\right) [\delta^{(j)}] . \tag{2.1}$$

Wir betrachten folgenden linearen Teilraum von $\mathcal{D}_m'$:

$$\mathcal{D}_{m+}' = \left\{ y \in \mathcal{D}_m' : y = \sum_{i=0}^{M} \beta_i [\delta^{(i)}] + [h],\ M \text{ nicht fest} \right\},$$

dabei ist $\beta_i = (\beta_{1i}, \dots, \beta_{mi})^\mathsf{T}$, $h(t) = \left(h_1(t), \dots, h_m(t)\right)^\mathsf{T}$ erfüllt die Voraussetzungen des Satzes 2.1 für $k = n$ (n Ordnung der betrachteten Differentialgleichung) und $h(t) = 0$ für $t < 0$. Die Elemente von $\mathcal{D}_{m+}'$ sind Distributionen mit Träger in $[0, \infty)$ (vgl. [6], S. 460). Für $y \in \mathcal{D}_{m+}'$ schreiben wir auch $y(t) = \sum_{i=0}^{M} \beta_i \delta^{(i)}(t) + h(t)$, für $t > 0$ ist $y(t) = h(t)$.

Gegeben seien der lineare Differentialoperator

$$Ly := A^{(n)} y^{(n)}(t) + \dots + A^{(1)} y'(t) + A^{(0)} y(t) ,$$

der Störvektor

$$\sum_{i=0}^{N} \alpha_i \delta^{(i)}(t) + k(t)$$

und beliebige linksseitige Anfangsbedingungen

$$y(-0) = y_0, \dots, y^{(n-1)}(-0) = y_0^{(n-1)} .$$

Hierbei ist

$$y^{(k)}(t) = \begin{pmatrix} y_1^{(k)}(t) \\ \vdots \\ y_m^{(k)}(t) \end{pmatrix}, \qquad y_0^{(k)} = \begin{pmatrix} y_{10}^{(k)} \\ \vdots \\ y_{m0}^{(k)} \end{pmatrix}, \qquad \alpha_i = \begin{pmatrix} \alpha_{1i} \\ \vdots \\ \alpha_{mi} \end{pmatrix}, \qquad k(t) = \begin{pmatrix} k_1(t) \\ \vdots \\ k_m(t) \end{pmatrix},$$

$$A^{(k)} = (a_{ij}^{(k)})_{i,j=1,\dots,m},\ a_{ij}^{(k)}, \alpha_{ji}, y_{j0}^{(k)} \in R^1 .$$

Unter Beachtung von Satz 2.1 treffen wir folgende Formulierung des *linksseitigen Anfangswertproblems*:

Problem [A⁻]: Gesucht ist eine Distribution $y \in \mathcal{D}_{m+}'$, die

$$A^{(n)} \tilde{D}^n y + \dots + A^{(1)} \tilde{D} y + A^{(0)} y = \sum_{i=0}^{N} \alpha_i [\delta^{(i)}] + [k] \tag{2.2}$$

mit

$$\tilde{D}^k y := \sum_{i=0}^{M} \beta_i [\delta^{(i+k)}] + [h^{(k)}] + \sum_{j=0}^{k-1} \left(h^{(k-j-1)}(+0) - y_0^{(k-j-1)}\right) [\delta^{(j)}] \tag{2.3}$$

erfüllt.

Für $u, v \in \mathcal{D}_{m+}'$ gilt

$$\tilde{D}^k(u + v) = \tilde{D}^k u + D^k v = D^k u + \tilde{D}^k v . \tag{2.4}$$

Das Problem [A⁻] ist offenbar äquivalent zu

1. $Lh = k(t)$ für $t > 0$ mit rechtsseitigen Anfangsbedingungen für $h(t)$; (2.5)

2. $$\sum_{k=0}^{n} A^{(k)} \left(\sum_{i=0}^{M} \beta_i[\delta^{(i+k)}] + \sum_{j=0}^{k-1} (h^{(k-j-1)}(+0) - y_0^{(k-j-1)} [\delta^{(j)}]\right) = \sum_{i=0}^{N} \alpha_i[\delta^{(i)}] . \quad (2.6)$$

Koeffizientenvergleich nach $[\delta^{(j)}]$ in (2.6) ergibt ein lineares Gleichungssystem für $\beta_0, \dots, \beta_M$ $(M > \mathrm{Max}(n, N))$ und $h(+0), \dots, h^{(n-1)}(+0)$. Ist das charakteristische Polynom $\Delta(p) \not\equiv 0$, dann sind durch (2.6) $\beta_0, \dots, \beta_M$ und von $h(+0), \dots, h^{(n-1)}(+0)$ gerade so viele rechtsseitige Grenzwerte eindeutig bestimmt, wie für die eindeutige Lösbarkeit von (2.5) ausreicht (vgl. auch [1], S. 24). Wir beweisen dies.

Satz 2.2 (Eindeutigkeitssatz): *Ist das charakteristische Polynom*

$$\Delta(p) = \det (A^{(n)}p^n + \dots + A^{(1)}p + A^{(0)})$$

nicht identisch Null, dann kann das Problem [A⁻] *höchstens eine Lösung haben.*

Beweis: Da sich jedes System n-ter Ordnung als System 1. Ordnung schreiben läßt und dabei das charakteristische Polynom erhalten bleibt (s. [1], (74)), können wir uns beim Beweis auf ein System 1. Ordnung beschränken. Es genügt zu zeigen (beachte (2.4)): Ist $y = \sum_{i=0}^{M} \beta_i[\delta^{(i)}] + [h] \in \mathcal{D}'_{m+}$ Lösung von

$$ADy + By = [0] , \qquad \det (Ap + B) \not\equiv 0 , \quad (2.7)$$

dann ist $y = [0]$.

1. Sei $A = 0$. Dann ist nach Voraussetzung $\det B \neq 0$, also existiert B^{-1}. Aus $By = [0]$ folgt dann $y = [0]$.

2. Sei $A \neq 0$, Rang $A = m$. Dann existiert A^{-1}. Aus (2.7) folgt

$$A \sum_{i=0}^{M} \beta_i[\delta^{(i+1)}] + Ah(+0) [\delta] + B \sum_{i=0}^{M} \beta_i[\delta^{(i)}] = [0] \quad (2.8)$$

und

$$Ah'(t) + Bh(t) = 0 \quad \text{für} \quad t > 0 . \quad (2.9)$$

Koeffizientenvergleich in (2.8) ergibt

$$\begin{aligned} A\beta_M &= 0 , \\ A\beta_{M-1} + B\beta_M &= 0 , \\ &\dots \\ A\beta_0 + B\beta_1 &= 0 , \\ Ah(+0) + B\beta_0 &= 0 . \end{aligned}$$

Hieraus ergibt sich der Reihe nach $\beta_M = 0, \dots, \beta_0 = 0$ und $h(+0) = 0$. Aus (2.9) und $h(+0) = 0$ folgt $h(t) = 0$ für $t > 0$.

3. Sei $A \neq 0$, Rang $A = r$, $1 \leqq r < m$. Dann gilt folgende Aussage Q:

(2.7) *ist in* $\mathcal{D}'_{m+}$ *nur trivial lösbar, wenn das folgende Problem* H

$$GD\hat{y} + H\hat{y} = [0] , \qquad \hat{y} \in \mathcal{D}'_{r+} , \qquad \det (Gp + H) \not\equiv 0 , \qquad 0 \leqq \mathrm{Rang}\, G \leqq r \quad (2.10)$$

nur trivial lösbar ist.

Da nach 1. und 2. das Problem H in den Randfällen Rang $G = 0$, Rang $G = r$ nur trivial lösbar ist, kommt man mit Hilfe der Aussage Q in endlich vielen Schritten auf eine einzelne Gleichung, deren charakteristisches Polynom nicht identisch Null ist, die also nach 1. und 2. nur trivial lösbar ist.

Der Beweis der Aussage Q geht wie folgt: Durch GAUSS-Algorithmus-Operationen werden $m - r$ Gleichungen in (2.7) von $Dy_1, \dots, Dy_m$ befreit. Mit Hilfe dieser Gleichungen und ihrer Derivierten kommen wir für den Teilvektor $\hat{y}$, bestehend aus r Komponenten von y, auf ein Problem H, das nach Voraussetzung nur trivial lösbar ist. Für die restlichen Komponenten gilt dann $Dy_{i_{r+1}} = [0], \dots, Dy_{i_m} = [0]$, woraus $y_{i_{r+1}} = [0], \dots, y_{i_m} = [0]$ folgt.

Entscheidend für die Anwendbarkeit der Methode der LAPLACE-Transformation zur Lösung des Problems [A⁻] ist der

Satz 2.3: *Es erfülle $f(t)$ die Voraussetzungen des Satzes 2.1, und $f^{(k)}(t)$ sei Laplace-transformierbar, dann gilt*

$$\mathcal{L}\{D^k[f]\} = p^k \mathcal{L}\{f(t)\} - \sum_{j=0}^{k-1} f^{(k-j-1)}(-0)\, p^j . \quad (2.11)$$

Beweis: Beachtet man $\mathcal{L}\{[f^{(k)}]\} = \mathcal{L}\{f^{(k)}(t)\}$,

$$\mathcal{L}\{[\delta^{(k)}]\} = p^k \quad \text{(s. [6], S. 309)} \tag{2.12}$$

und den Differentiationssatz der klassischen LAPLACE-Transformation (s. [6], S. 305), dann folgt aus (2.1)

$$\mathcal{L}\{D^k[f]\} = p^k\mathcal{L}\{f(t)\} - \sum_{j=0}^{k-1} f^{(k-j-1)}(+0)\, p^j + \sum_{j=0}^{k-1} \left(f^{(k-j-1)}(+0) - f^{(k-j-1)}(-0)\right) p^j$$

$$= p^k\mathcal{L}\{f(t)\} - \sum_{j=0}^{k-1} f^{(k-j-1)}(-0)\, p^j .$$

Bezeichnet $s(t)$ die HEAVISIDEsche Sprungfunktion und $[sf]$ die von $s(t)\, f(t)$ erzeugte Distribution, dann ergibt sich aus (2.11) folgende für die Rücktransformation wichtige Formel

$$\mathcal{L}\{D^k[sf]\} = p^k\mathcal{L}\{f(t)\} . \tag{2.13}$$

Für das Aufstellen der Bildgleichung von (2.2) benötigen wir noch den folgenden Satz, der wie Satz 2.3 bewiesen wird.

Satz 2.4: *Sei* $y = \sum_{i=0}^{M} \beta_i[\delta^{(i)}] + [h] \in \mathcal{D}'_{m+}$ *und* $h^{(n)}(t)$ *Laplace-transformierbar, dann gilt*

$$\mathcal{L}\{\tilde{D}^k y\} = p^k\mathcal{L}\{y\} - \sum_{j=0}^{k-1} y_0^{(k-j-1)} p^j , \qquad k = 1, \dots, n . \tag{2.14}$$

Zur Lösung des Problems $[\mathrm{A}^-]$ stellt man mit Hilfe von (2.14) das Bildsystem zu (2.2) auf, dieses ist

$$(A^{(n)}p^n + \dots + A^{(1)}p + A^{(0)})\, \mathcal{L}\{y\} = \sum_{k=1}^{n} \sum_{j=0}^{k-1} A^{(k)} y_0^{(k-j-1)} p^j + \sum_{i=0}^{N} \alpha_i p^i + \mathcal{L}\{k(t)\} . \tag{2.15}$$

Ist $\Delta(p) \not\equiv 0$, existiert $(A^{(n)}p^n + \dots + A^{(1)}p + A^{(0)})^{-1} =: G(p)^{\mathrm{T}}$, und man erhält aus (2.15)

$$\mathcal{L}\{y\} = Y(p) = \sum_{k=1}^{n} \sum_{j=0}^{k-1} G(p)^{\mathrm{T}} A^{(k)} y_0^{(k-j-1)} p^j + \sum_{i=0}^{N} G(p)^{\mathrm{T}} \alpha_i p^i + G(p)^{\mathrm{T}} \mathcal{L}\{k(t)\} . \tag{2.16}$$

Die Elemente $G_{ik}(p)$ von $G(p)$ sind rationale Funktionen

$$G_{ik}(p) = \frac{r_{ik}(p)}{s_{ik}(p)}, \qquad \mathrm{Grad}\; s_{ik}(p) - \mathrm{Grad}\; r_{ik}(p) =: n_{ik} . \tag{2.17}$$

Ist das System normal, d. h. det $A^{(n)} \neq 0$, dann ist $n_{ik} \geqq n$.

Mit Hilfe von (2.12) und dem Faltungssatz (s. [6], S. 306) kann man (2.16) zurücktransformieren. Beachtet man noch den Differentiationssatz für die Faltung (s. [6], S. 305), erhält man den

Satz 2.5: *Sei* det $A^{(n)} \neq 0$ *und* $k(t)$ *für* $t > 0$ *stetig und* $\mathcal{L}$*-transformierbar. Dann hat das Problem* $[\mathrm{A}^-]$ *eine eindeutige Lösung* $y \in \mathcal{D}'_{m+}$, *die mit der Methode der Laplace-Transformation gefunden werden kann.*

Ist das System anormal, dann wird bei $n_{ik} < 0$ für die Rücktransformation noch (2.13) benötigt. Damit y die Voraussetzungen des Satzes 2.4 erfüllt, sind stärkere Differenzierbarkeitsvoraussetzungen an k zu stellen. Es gilt der

Satz 2.6: *Sei* det $A^{(n)} = 0$, $\Delta(p) = \det (A^{(n)}p^n + \dots + A^{(1)}p + A^{(0)}) \not\equiv 0$, $N_i = \min\limits_{1 \leqq k \leqq n} n_{ik}$ (vgl. (2.17)), $M_i = \max(0, n - N_i)$. *Ist* $k_i^{(M_i)}(t)$ *für* $t > 0$ *stetig und* $\mathcal{L}$*-transformierbar* $(i = 1, \dots, m)$, *dann hat das Problem* $[\mathrm{A}^-]$ *eine eindeutige Lösung* $y \in \mathcal{D}'_{m+}$, *die mit der Methode der Laplace-Transformation gefunden werden kann.*

In Vorbereitung auf die nächsten Abschnitte zerlegen wir das Problem $[\mathrm{A}^-]$ in folgende Teilprobleme:

Problem $[\mathrm{A}_\delta^-]$: $\quad A^{(n)}D^n y + \dots + A^{(1)}Dy + A^{(0)}y = \sum_{i=0}^{N} \alpha_i[\delta^{(i)}] , \qquad y \in \mathcal{D}'_{m+} ,$

Problem $[\mathrm{A}_{\mathrm{I}}^-]$: $\quad A^{(n)}\tilde{D}^n y + \dots + A^{(1)}\tilde{D}y + A^{(0)}y = [0] , \qquad y \in \mathcal{D}'_{m+} ,$

Problem $[\mathrm{A}_{\mathrm{II}}^-]$: $\quad A^{(n)}D^n y + \dots + A^{(1)}Dy + A^{(0)}y = [k] , \qquad y \in \mathcal{D}'_{m+} .$

Die Lösung des Problems $[\mathrm{A}^-]$ im allgemeinen Fall setzt sich offenbar additiv aus den Lösungen der Probleme $[\mathrm{A}_\delta^-]$, $[\mathrm{A}_{\mathrm{I}}^-]$ und $[\mathrm{A}_{\mathrm{II}}^-]$ zusammen (beachte (2.4)).

3. Approximationssatz für das Problem $[A_\delta^-]$

Unter der einzigen Voraussetzung $\Delta(p) \not\equiv 0$ hat nach Satz 2.5 bzw. 2.6 das Problem $[A_\delta^-]$ eine eindeutige Lösung $y \in \mathcal{D}'_{m+}$. Wir approximieren $[\delta^{(i)}]$ durch $h_\eta^{(i)}(t)$, wobei $h_\eta(t)$ folgende Eigenschaften hat: $h_\eta(t) \geqq 0$, $h_\eta(t) = 0$ für $t \notin [0, \eta]$, $\int\limits_{-\infty}^{+\infty} h_\eta(t)\,dt = \int\limits_0^\eta h_\eta(t)\,dt = 1$, $h_\eta(t)$ hat stetige Ableitungen bis zur $(i-1)$-ten Ordnung und die i-te Ableitung ist stückweise stetig. Zusätzlich verlangen wir, daß $k(t) = \sum\limits_{i=0}^N \alpha_i h_\eta^{(i)}(t)$ die Voraussetzungen des Satzes 2.5 bzw. 2.6 erfüllt. Dann hat

$$A^{(n)}D^n y + \dots + A^{(1)}Dy + A^{(0)}y = \sum_{i=0}^N \alpha_i[h_\eta^{(i)}] \tag{3.1}$$

eine eindeutige Lösung $y_\eta \in \mathcal{D}'_{m+}$. Dann gilt der

Satz 3.1: *Sei $y \in \mathcal{D}'_{m+}$ die Lösung des Problems $[A_\delta^-]$ und $y_\eta \in \mathcal{D}'_{m+}$ die Lösung von* (3.1). *Dann gilt*

$$\lim_{\eta\to+0} y_\eta(t) = y(t) \quad \textit{für alle } t > 0\,.$$

Beweis: Es sei

$$G_{ik}(p) = R_{ik}(p) + \sum_{j=0}^r \beta_j^{ik} p^j\,, \qquad r < n \cdot m\,,$$

wobei $R_{ik}(p)$ eine echt gebrochen rationale Funktion ist und

$$r_{ik} = \mathcal{L}^{-1}\{R_{ik}(p)\}\,.$$

Die Methode der $\mathcal{L}$-Transformation liefert für die s-te Komponente y_s von y

$$y_s = \sum_{i=0}^N \sum_{l=1}^m \alpha_{li}[s r_{ls}^{(i)}] + \sum_{i=1}^N \sum_{l=1}^m \alpha_{li} \sum_{j=0}^{i-1} r_{ls}^{(i-j-1)}(+0)\,[\delta^{(j)}] + \sum_{i=0}^N \sum_{l=1}^m \sum_{t=0}^r \beta_t^{ls}\alpha_{li}[\delta^{(i+t)}] \tag{3.2}$$

und für die s-te Komponente y_s^η von y_η

$$y_s^\eta = \sum_{i=0}^N \sum_{l=1}^m \alpha_{li}[r_{ls}(t) * h_\eta^{(i)}(t)] + \sum_{i=0}^N \sum_{l=1}^m \sum_{t=0}^r \beta_t^{ls}\alpha_{li} D^t[h_\eta^{(i)}] \tag{3.3}$$

Für $t > \eta > 0$ folgt aus (3.2) und (3.3)

$$y_s(t) = \sum_{i=0}^N \sum_{l=1}^m \alpha_{li} r_{ls}^{(i)}(t)\,, \qquad y_s^\eta(t) = \sum_{i=0}^N \sum_{l=1}^m \alpha_{li} r_{ls}(t) * h_\eta^{(i)}(t)\,.$$

Nun ist

$$\lim_{\eta\to+0} r_{ls}(t) * h_\eta^{(i)}(t) = \lim_{\eta\to+0} \int_0^\eta r_{ls}(t-\tau)\, h_\eta^{(i)}(\tau)\,d\tau = \lim_{\eta\to+0} \int_0^\eta r_{ls}^{(i)}(t-\tau)\, h_\eta(\tau)\,d\tau$$

$$= \lim_{\eta\to+0} r_{ls}^{(i)}(t-\vartheta\eta) \int_0^\eta h_\eta(\tau)\,d\tau = \lim_{\eta\to+0} r_{ls}^{(i)}(t-\vartheta\eta) = r_{ls}^{(i)}(t) \quad , \quad 0 < \vartheta < 1\,.$$

Damit ist der Satz bewiesen.

Beispiel 3.1: $y_1' + y_2 = 0$,

$$y_1 + y_2 = \delta'(t)\,.$$

Anfangsbedingungen: $y_1(-0) = 0$, $y_2(-0) = 0$.
Nach (2.14) folgt für das Bildsystem

$$pY_1(p) + Y_2(p) = 0\,,$$

$$Y_1(p) + Y_2(p) = p\,.$$

Es ist

$$\Delta(p) = \begin{vmatrix} p & 1 \\ 1 & 1 \end{vmatrix} = p - 1\,.$$

Man erhält

$$Y_1(p) = -1 - \frac{1}{p-1}, \qquad Y_2(p) = p + 1 + \frac{1}{p-1}$$

und hieraus

$$y_1(t) = -\delta(t) - e^t\,, \qquad y_2(t) = \delta'(t) + \delta(t) + e^t\,, \qquad t \geqq 0\,.$$

Wir approximieren nun $\delta'(t)$ und erhalten das Problem

$$(y_1^\eta)' + y_2^\eta = 0\,,$$

$$y_1^\eta + y_2^\eta = h_\eta'(t) \quad \text{mit} \quad h_\eta(t) = \frac{30}{\eta^6}\, t(t-\eta)^4 \big(s(t) - s(t-\eta)\big)\,.$$

Anfangsbedingungen: $y_1^\eta(-0) = 0$, $y_2^\eta(-0) = 0$.

Da das System anormal ist, brauchen die rechtsseitigen Grenzwerte nicht Null zu sein. Man erhält mit der Methode der Laplace-Transformation für $t \geqq 0$ die Lösung

$$y_1^\eta(t) = \frac{30}{\eta^6}\left[A + Bt + C\frac{t^2}{2} + D\frac{t^3}{6} + 5t^4 - A\,\mathrm{e}^t\right] -$$

$$-\frac{30}{\eta^6}\left[E\left(1 + (t-\eta) + \frac{(t-\eta)^2}{2} + \frac{(t-\eta)^3}{6}\right) + 5(t-\eta)^4 - E\,\mathrm{e}^{t-\eta}\right] s(t-\eta),$$

$$y_2^\eta(t) = -\frac{30}{\eta^6}\left[B + Ct + D\frac{t^2}{2} + 20t^3 - A\,\mathrm{e}^t\right] +$$

$$+\frac{30}{\eta^6}\left[E\left(1 + (t-\eta) + \frac{(t-\eta)^2}{2}\right) + 20(t-\eta)^3 - E\,\mathrm{e}^{t-\eta}\right] s(t-\eta)$$

mit

$$A = 120 - 96\eta + 36\eta^2 - 8\eta^3 + \eta^4, \qquad B = 120 - 96\eta + 36\eta^2 - 8\eta^3,$$

$$C = 120 - 96\eta + 36\eta^2, \qquad D = 120 - 96\eta, \qquad E = 120 + 24\eta.$$

Es ist

$$\lim_{t\to+0} y_1^\eta(t) = 0, \qquad \lim_{t\to+0} y_2^\eta(t) = \frac{30}{\eta^2},$$

und für $t > \eta$ gilt

$$y_1^\eta(t) = \frac{30}{\eta^6}(E\,\mathrm{e}^{-\eta} - A)\,\mathrm{e}^t = \left(-1 - 6!\sum_{n=1}^{\infty}\frac{n+1}{(n+6)!}(-\eta)^n\right)\mathrm{e}^t,$$

$$y_2^\eta(t) = -y_1^\eta(t).$$

Abbildung 1 zeigt $y_\eta(t)$ für verschiedene Werte von η zusammen mit der Grenzlösung $y(t)$ für $t > 0$.

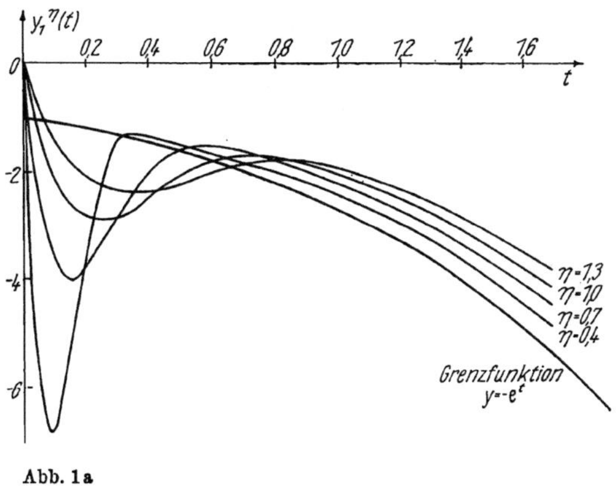

Abb. 1a

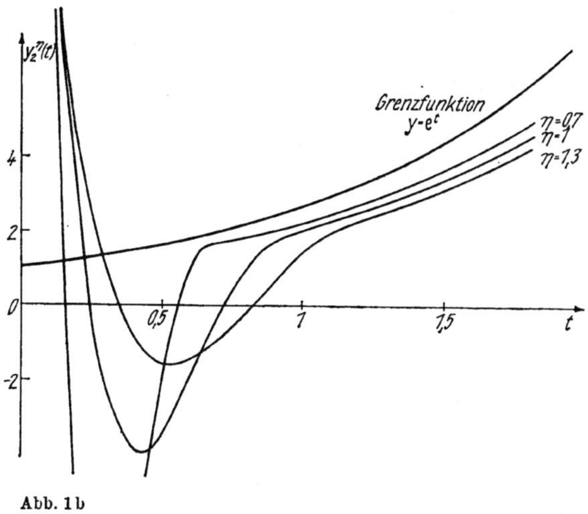

Abb. 1b

4. Approximationssatz für das Problem $[\mathrm{A}_\mathrm{I}^-]$

Wir betrachten in diesem Abschnitt das Problem $[\mathrm{A}_\mathrm{I}^-]$ im anormalen Fall, d. h. $\det A^{(n)} = 0$, und es sei $\Delta(p) = \det(A^{(n)}p^n + \dots + A^{(1)}p + A^{(0)}) \not\equiv 0$. Dabei ist nicht ausgeschlossen, daß $A^{(n)}$ und auch weitere Koeffizientenmatrizen $A^{(j)}$ gleich der Nullmatrix sind. Nach Satz 2.6 hat das Problem $[\mathrm{A}_\mathrm{I}^-]$ eine eindeutige Lösung $y \in \mathcal{D}'_{m+}$. Daneben betrachten wir ein gestörtes Problem $\mathrm{A}_\mathrm{I}^\varepsilon$:

Differentialgleichungssystem:

$$L_\varepsilon y_\varepsilon := A_\varepsilon^{(n)} y_\varepsilon^{(n)}(t) + \dots + A_\varepsilon^{(1)} y_\varepsilon'(t) + A_\varepsilon^{(0)} y_\varepsilon(t) = 0. \tag{4.1}$$

Anfangsbedingungen:

$$y_\varepsilon(-0) = y_0, \dots, y_\varepsilon^{(n-1)}(-0) = y_0^{(n-1)}, \tag{4.2}$$

$$\det A_\varepsilon^{(n)} \neq 0 \quad \text{für} \quad 0 < \varepsilon < \varepsilon_0,$$

$$A_\varepsilon^{(k)} = A^{(k)} + \varepsilon \hat{A}_\varepsilon^{(k)}, \quad \text{wobei } \hat{a}_{\varepsilon ij}^{(k)} \text{ Polynome in } \varepsilon \text{ sind},$$

$$\Delta_\varepsilon(p) := \det(A_\varepsilon^{(n)} p^n + \dots + A_\varepsilon^{(1)} p + A_\varepsilon^{(0)}). \tag{4.3}$$

Das Problem A_I^ε hat eine klassische Lösung $y_\varepsilon(t)$, für die gilt

$$y_\varepsilon(-0) = y_\varepsilon(+0) = y_0, \dots, y_\varepsilon^{(n-1)}(-0) = y_\varepsilon^{(n-1)}(+0) = y_0^{(n-1)}.$$

Es ist nun die Frage, unter welchen Bedingungen $y_\varepsilon(t)$ für $t > 0$ bei $\varepsilon \to +0$ gegen $y(t)$ konvergiert. Bevor wir den entsprechenden Satz formulieren, führen wir noch einige Bezeichnungen ein.

Mit $p_1, \dots, p_{r_0}$ bezeichnen wir die verschiedenen Nullstellen von $\Delta(p)$ $(0 \leqq \text{Grad}\, \Delta(p) = n_0 < n \cdot m)$ mit den Vielfachheiten $n_1, \dots, n_{r_0}$ $(n_1 + \dots + n_{r_0} = n_0)$. Durch die ε-Störung spaltet p_k, $k = 1, \dots, r_0$, auf in m_k Nullstellen $p_{k1}(\varepsilon), \dots, p_{km_k}(\varepsilon)$ mit den Vielfachheiten $m_{k1}, \dots, m_{km_k}(m_{k1} + \dots + m_{km_k} = n_k)$. Es gibt eine Zahl $\delta > 0$, so daß

$$p_{kj}(\varepsilon) \neq p_{kl}(\varepsilon), \qquad j \neq l, \quad \text{für} \quad 0 < |\varepsilon| < \delta.$$

Es gilt

$$\lim_{\varepsilon \to +0} p_{kj}(\varepsilon) = p_k, \qquad j = 1, \dots, m_k, \qquad k = 1, \dots, r_0.$$

Die restlichen Nullstellen von $\Delta_\varepsilon(p)$ bezeichnen wir mit $p_{n_0+1}(\varepsilon), \dots, p_{n \cdot m}(\varepsilon)$. Diese brauchen nicht alle verschieden zu sein.

Satz 4.1: *Sei $y \in \mathcal{D}'_{m+}$ die Lösung des Problems $[A_I^-]$ und $y_\varepsilon(t)$ die Lösung des gestörten Problems A_I^ε. Falls*

$$\lim_{\varepsilon \to +0} \operatorname{Re} p_j(\varepsilon) = -\infty, \qquad j = n_0 + 1, \dots, n \cdot m, \tag{4.4}$$

dann gilt

$$\lim_{\varepsilon \to +0} y_\varepsilon(t) = y(t) \quad \textit{für alle } t > 0.$$

Der Beweis des Satzes 4.1 stützt sich auf die folgenden Hilfssätze.

Hilfssatz 4.1: *Sei $f(z)$ im einfach zusammenhängenden Gebiet G analytisch und $L \subset G$ eine glatte, geschlossene, mathematisch positiv orientierte Kurve, $z_1, \dots, z_r$ innerhalb L gelegene Punkte und $P(z)$ das durch die Bedingungen*

$$P(z_k) = f(z_k), \dots, P^{(m_k-1)}(z_k) = f^{(m_k-1)}(z_k), \qquad k = 1, \dots, r,$$

eindeutig bestimmte Hermitesche Interpolationspolynom. Dann ist für z innerhalb L

$$P(z) = \frac{1}{2\pi i} \int\limits_L \frac{\omega(\zeta) - \omega(z)}{\omega(\zeta)\,(\zeta - z)} f(\zeta)\, d\zeta \tag{4.5}$$

mit

$$\omega(z) = (z - z_1)^{m_1} \cdot \dots \cdot (z - z_r)^{m_r}.$$

Diesen Hilfssatz findet man im Spezialfall des LAGRANGEschen Interpolationspolynoms $(m_1 = \dots = m_r = 1)$ in [12], S. 139. Der Beweis von (4.5) ergibt sich leicht durch Anwendung des Residuensatzes und durch Berechnung der Residuen aus der LAURENT-Reihe. Man erhält

$$\frac{1}{2\pi i} \int\limits_L \frac{\omega(\zeta) - \omega(z)}{\omega(\zeta)\,(\zeta - z)} f(\zeta)\, d\zeta = \sum_{i=1}^{r} \operatorname{Res}\left[\frac{\omega(\zeta) - \omega(z)}{\omega(\zeta)\,(\zeta - z)} f(\zeta),\ \zeta = z_i\right]$$

$$= \sum_{i=1}^{r} \sum_{l=0}^{m_i-1} \frac{1}{l!} \frac{\omega(z)}{(z - z_i)^{m_i - l}} \sum_{k=0}^{m_i-l-1} \frac{1}{k!} \left[\frac{(z - z_i)^{m_i}}{\omega(z)}\right]^{(k)}_{z=z_i} (z - z_i)^k f^{(l)}(z_i).$$

Das ist genau das HERMITEsche Interpolationspolynom (s. [12], S. 120). Aus der CAUCHYschen Integralformel ergibt sich nun sofort der Fehler der Interpolationsformel, denn es ist für z innerhalb L

$$f(z) = \frac{1}{2\pi i} \int\limits_L \frac{f(\zeta)}{\zeta - z}\, d\zeta = \frac{1}{2\pi i} \int\limits_L \frac{\omega(\zeta) - \omega(z)}{\omega(\zeta)\,(\zeta - z)} f(\zeta)\, d\zeta + \frac{\omega(z)}{2\pi i} \int\limits_L \frac{f(\zeta)}{\omega(\zeta)\,(\zeta - z)}\, d\zeta$$

also

$$f(z) = P(z) + R(z) \quad \text{mit} \quad R(z) = \frac{\omega(z)}{2\pi i} \int\limits_L \frac{f(\zeta)}{\omega(\zeta)\,(\zeta - z)}\, d\zeta. \tag{4.6}$$

Hilfssatz 4.2: *Sei $g(p)$ eine ganze Funktion, dann gilt*

$$\lim_{\varepsilon \to +0} \sum_{\alpha=1}^{m_k} \frac{1}{(m_{k\alpha} - 1)!} \left[\frac{(p - p_{k\alpha}(\varepsilon))^{m_{k\alpha}}}{\Delta_\varepsilon(p)} g(p)\right]^{(m_{k\alpha}-1)}_{p = p_{k\alpha}(\varepsilon)} = \frac{1}{(n_k - 1)!} \left[\frac{(p - p_k)^{n_k}}{\Delta(p)} g(p)\right]^{(n_k-1)}_{p = p_k}. \tag{4.7}$$

Beweis: Es ist

$$\Delta(p) = a_0 r(p)\,(p - p_k)^{n_k}, \qquad r(p_k) \neq 0,$$

und

$$\Delta_\varepsilon(p) = a_\varepsilon r_\varepsilon(p)\,\big(p - p_{k1}(\varepsilon)\big)^{m_{k1}} \cdot \dots \cdot \big(p - p_{km_k}(\varepsilon)\big)^{m_{km_k}}, \qquad r_\varepsilon(p_{kj}) \neq 0, \qquad j = 1, \dots, m_k.$$

Nun ist

$$\sum_{a=1}^{m_k} \frac{1}{(m_{ka}-1)!} \left[\frac{(p-p_{ka}(\varepsilon))^{m_{ka}}}{\Delta_\varepsilon(p)} g(p)\right]^{(m_{ka}-1)}_{p=p_{ka}(\varepsilon)}$$

$$= \sum_{a=1}^{m_k} \frac{1}{(m_{ka}-1)!} \left[\prod_{\substack{j=1\\ j\neq a}}^{m_k} \frac{1}{(p-p_{kj}(\varepsilon))^{m_{kj}}} \frac{1}{a_0 r(p)} g(p)\right]^{(m_{ka}-1)}_{p=p_{ka}(\varepsilon)} +$$

$$+ \sum_{a=1}^{m_k} \frac{1}{(m_{ka}-1)!} \left[\prod_{\substack{j=1\\ j\neq a}}^{m_k} \frac{1}{(p-p_{kj}(\varepsilon))^{m_{kj}}} \left(\frac{1}{a_\varepsilon r_\varepsilon(p)} - \frac{1}{a_0 r(p)}\right) g(p)\right]^{(m_{ka}-1)}_{p=p_{ka}(\varepsilon)}. \tag{4.8}$$

Wir betrachten

$$f(p) = \frac{(p-p_k)^{n_k}}{\Delta(p)} g(p) = \frac{1}{a_0 r(p)} g(p)$$

und das HERMITEsche Interpolationspolynom $P(z)$ mit

$$P(p_{ka}(\varepsilon)) = f(p_{ka}(\varepsilon)), \dots, P^{(m_{ka}-1)}(p_{ka}(\varepsilon)) = f^{(m_{ka}-1)}(p_{ka}(\varepsilon)), \qquad a = 1, \dots, m_k .$$

Dann ist nach (4.6)

$$f(p) = P(p) + R(p) \tag{4.9}$$

mit

$$R(p) = \frac{\omega(p)}{2\pi i} \int\limits_L \frac{f(\zeta)}{\omega(\zeta)(\zeta-p)} d\zeta, \qquad \omega(p) = (p-p_{k1}(\varepsilon))^{m_{k1}} \cdot \ldots \cdot (p-p_{km_k}(\varepsilon))^{m_{km_k}} .$$

Dabei ist L so zu wählen, daß von den Nullstellen von $\Delta(p)$ und $\Delta_\varepsilon(p)$ genau $p_k, p_{k1}(\varepsilon), \dots, p_{km_k}(\varepsilon)$ innerhalb L liegen.

Bilden wir von (4.9) die (n_k-1)-te Ableitung, dann ist $P^{(n_k-1)}(p) = \text{const.}$, und man erhält (vgl. [12], S. 129)

$$\sum_{a=1}^{m_k} \frac{1}{(m_{ka}-1)!} \left[\prod_{\substack{j=1\\ j\neq a}}^{m_k} \frac{1}{(p-p_{kj}(\varepsilon))^{m_{kj}}} \frac{1}{a_0 r(p)} g(p)\right]^{(m_{ka}-1)}_{p=p_{ka}(\varepsilon)} = \frac{1}{(n_k-1)!} f^{(n_k-1)}(p) - \frac{1}{(n_k-1)!} R^{(n_k-1)}(p). \tag{4.10}$$

Setzen wir in (4.10) $p = p_k$, dann ist $\lim\limits_{\varepsilon\to+0} R^{(n_k-1)}(p_k) = 0$, so daß nach Grenzübergang $\varepsilon \to +0$ die rechten Seiten von (4.10) und (4.7) übereinstimmen.

Jetzt führen wir dieselbe Überlegung mit

$$f_\varepsilon(p) = \left(\frac{1}{a_\varepsilon r_\varepsilon(p)} - \frac{1}{a_0 r(p)}\right) g(p)$$

durch und erhalten

$$\sum_{a=1}^{m_k} \frac{1}{(m_{ka}-1)!} \left[\prod_{\substack{j=1\\ j\neq a}}^{m_k} \frac{1}{(p-p_{kj}(\varepsilon))^{m_{kj}}} \left(\frac{1}{a_\varepsilon r_\varepsilon(p)} - \frac{1}{a_0 r(p)}\right) g(p)\right]^{(m_{ka}-1)}_{p=p_{ka}(\varepsilon)}$$

$$= \frac{1}{(n_k-1)!} f_\varepsilon^{(n_k-1)}(p) - \frac{1}{(n_k-1)!} R_\varepsilon^{(n_k-1)}(p). \tag{4.11}$$

Setzen wir in (4.11) $p = p^* \neq p_k$, dann ist

$$\lim_{\varepsilon\to+0} f_\varepsilon^{(n_k-1)}(p^*) = 0 \quad \text{und} \quad \lim_{\varepsilon\to+0} R_\varepsilon^{(n_k-1)}(p^*) = 0 .$$

Damit ist der Hilfssatz 4.2 bewiesen.

Beweis von Satz 4.1: Setzen wir $\sum\limits_{k=1}^{n} \sum\limits_{j=0}^{k-1} A^{(k)} y_0^{(k-j-1)} p^j = \sum\limits_{u=0}^{n-1} \gamma_u p^u$ und $A_{ik}(p) = \Delta(p)\, G_{ik}(p)$, dann ist nach (2.16)

$$\mathcal{L}\{y_s\} = Y_s(p) = \frac{1}{\Delta(p)} \sum_{l=1}^{m} \sum_{u=0}^{n-1} A_{ls}(p)\, \gamma_{lu} p^u . \tag{4.12}$$

Setzen wir $\sum\limits_{k=1}^{n} \sum\limits_{j=0}^{k-1} A_\varepsilon^{(k)} y_0^{(k-j-1)} p^j = \sum\limits_{u=0}^{n-1} \gamma_u^\varepsilon p^u$ und bezeichnen mit $A_{ik}^\varepsilon(p)$ die Adjunkte der i-ten Zeile und k-ten Spalte von $A_\varepsilon^{(n)} p^n + \dots + A_\varepsilon^{(1)} p + A_\varepsilon^{(0)}$, dann ist nach (2.16)

$$\mathcal{L}\{y_s^\varepsilon\} = Y_s^\varepsilon(p) = \frac{1}{\Delta_\varepsilon(p)} \sum_{l=1}^{m} \sum_{u=0}^{n-1} A_{ls}^\varepsilon(p)\, \gamma_{lu}^\varepsilon p^u . \tag{4.13}$$

Nun ist auf Grund der Voraussetzung

$$A_{ik}^{\varepsilon}(p) = A_{ik}(p) + \varepsilon \hat{A}_{ik}^{\varepsilon}(p)\,, \tag{4.14}$$

$$\gamma_u^{\varepsilon} = \gamma_u + \varepsilon \hat{\gamma}_u^{\varepsilon}\,, \tag{4.15}$$

wobei $\hat{A}_{ik}^{\varepsilon}(p)$ Polynome in p sind, deren Koeffizienten ebenso wie $\hat{\gamma}_u^{\varepsilon}$ Polynome in ε sind.

Die Partialbruchzerlegung ergibt

$$\frac{1}{\Delta_\varepsilon(p)} = S_\varepsilon(p) + Q_\varepsilon(p)\,, \tag{4.16}$$

wobei

$$S_\varepsilon(p) = \sum_{k=1}^{r_0} \sum_{a=1}^{m_k} \sum_{q=1}^{m_{ka}} \frac{d_{kaq}^{\varepsilon}}{(p - p_{ka}(\varepsilon))^q}$$

mit

$$d_{kaq}^{\varepsilon} = \frac{1}{(m_{ka} - q)!} \frac{d^{m_{ka}-q}}{dp^{m_{ka}-q}} \left[\frac{(p - p_{ka}(\varepsilon))^{m_{ka}}}{\Delta_\varepsilon(p)}\right]_{p = p_{ka}(\varepsilon)}$$

ist und in $Q_\varepsilon(p)$ die Nullstellen $p_{n_0+1}(\varepsilon), \dots, p_{n\cdot m}(\varepsilon)$ stecken. Aus (4.13) folgt nun mit (4.14), (4.15), (4.16)

$$\begin{aligned} Y_s^{\varepsilon}(p) = {} & S_\varepsilon(p) \sum_{l=1}^{m} \sum_{u=0}^{n-1} A_{ls}(p)\, \gamma_{lu} p^u + \varepsilon S_\varepsilon(p) \sum_{l=1}^{m} \sum_{u=0}^{n-1} \hat{A}_{ls}^{\varepsilon}(p)\, \gamma_{lu} p^u + \\ & + \varepsilon S_\varepsilon(p) \sum_{l=1}^{m} \sum_{u=0}^{n-1} A_{ls}(p)\, \hat{\gamma}_{lu}^{\varepsilon} p^u + \varepsilon^2 S_\varepsilon(p) \sum_{l=1}^{m} \sum_{u=0}^{n-1} \hat{A}_{ls}^{\varepsilon}(p)\, \hat{\gamma}_{lu}^{\varepsilon} p^u \\ & + Q_\varepsilon(p) \sum_{l=1}^{m} \sum_{u=0}^{n-1} A_{ls}(p)\, \gamma_{lu} p^u + \varepsilon Q_\varepsilon(p) \sum_{l=1}^{m} \sum_{u=0}^{n-1} \hat{A}_{ls}^{\varepsilon}(p)\, \gamma_{lu} p^u \\ & + \varepsilon Q_\varepsilon(p) \sum_{l=1}^{m} \sum_{u=0}^{n-1} A_{ls}(p)\, \hat{\gamma}_{lu}^{\varepsilon} p^u + \varepsilon^2 Q_\varepsilon(p) \sum_{l=1}^{m} \sum_{u=0}^{n-1} \hat{A}_{ls}^{\varepsilon}(p)\, \hat{\gamma}_{lu}^{\varepsilon} p^u\,. \end{aligned} \tag{4.17}$$

Die Rücktransformation der letzten vier Ausdrücke ergibt δ-Distributionen und ihre Ableitungen und Ausdrücke der Form

$$r(\varepsilon, p_{11}(\varepsilon), \dots, p_{n\cdot m}(\varepsilon))\, \frac{1}{(\alpha - 1)!}\, t^{\alpha-1}\, e^{p_i(\varepsilon)\, t}\,, \qquad i = n_0 + 1, \dots, n \cdot m\,, \tag{4.18}$$

wobei $r(\varepsilon, p_{11}(\varepsilon), \dots, p_{n\cdot m}(\varepsilon))$ eine rationale Funktion ihrer Argumente ist. Die Wurzeln $p_{11}(\varepsilon), \dots, p_{n\cdot m}(\varepsilon)$ gestatten Puiseux-Entwicklungen mit endlich vielen negativen Exponenten. Auf Grund der Voraussetzung (4.4) konvergieren die Ausdrücke (4.18) für jedes feste $t > 0$ bei $\varepsilon \to +0$ gegen Null.

Wir zeigen jetzt, daß für festes $t > 0$ gilt

$$\lim_{\varepsilon \to +0} \mathcal{L}^{-1}\{S_\varepsilon(p)\, p^j\} = \mathcal{L}^{-1}\left\{\frac{1}{\Delta(p)}\, p^j\right\}, \qquad j \text{ beliebig fest}\,. \tag{4.19}$$

Dann gehen wegen des ε-Faktors im 2. bis 4. Ausdruck in (4.17) die rücktransformierten Ausdrücke für $t > 0$ bei $\varepsilon \to +0$ gegen Null. Vergleicht man den ersten Ausdruck von (4.17) mit (4.12), dann folgt aus (4.19) die Behauptung des Satzes. Bei der Rücktransformation aller 8 Ausdrücke in (4.17) können δ-Distributionen im Spiel sein, die sich aber aufheben müssen, da nach Voraussetzung das System (4.1) normal ist.

Wir kommen nun zum Beweis von (4.19): Die Partialbruchzerlegung von $1/\Delta(p)$ ist

$$\frac{1}{\Delta(p)} = \sum_{k=1}^{r_0} \sum_{v=1}^{n_k} \frac{c_{kv}}{(p - p_k)^v} \tag{4.20}$$

mit

$$c_{kv} = \frac{1}{(n_k - v)!} \frac{d^{n_k - v}}{dp^{n_k - v}} \left[\frac{(p - p_k)^{n_k}}{\Delta(p)}\right]_{p = p_k}.$$

Danach ist (4.19) bewiesen, wenn für $t > 0$ gilt

$$\lim_{\varepsilon \to +0} \mathcal{L}^{-1}\left\{\sum_{a=1}^{m_k} \sum_{q=1}^{m_{ka}} d_{kaq}^{\varepsilon} \frac{p^j}{(p - p_{ka}(\varepsilon))^q}\right\} = \mathcal{L}^{-1}\left\{\sum_{v=1}^{n_k} c_{kv} \frac{p^j}{(p - p_k)^v}\right\}. \tag{4.21}$$

Nun ist

$$\begin{aligned} \sum_{a=1}^{m_k} \sum_{q=1}^{m_{ka}} d_{kaq}^{\varepsilon} \frac{p^j}{(p - p_{ka}(\varepsilon))^q} = {} & \sum_{a=1}^{m_k} \sum_{q=1}^{m_{ka}} \sum_{l=0}^{q-1} d_{kaq}^{\varepsilon} \binom{j}{l} p_{ka}(\varepsilon)^{j-l} (p - p_{ka}(\varepsilon))^{l-q} + \\ & + \sum_{a=1}^{m_k} \sum_{q=1}^{m_{ka}} \sum_{l=q}^{j} d_{kaq}^{\varepsilon} \binom{j}{l} p_{ka}(\varepsilon)^{j-l} (p - p_{ka}(\varepsilon))^{l-q} \end{aligned} \tag{4.22}$$

und

$$\sum_{v=1}^{n_k} c_{kv} \frac{p^j}{(p - p_k)^v} = \sum_{v=1}^{n_k} \sum_{l=0}^{v-1} c_{kv} \binom{j}{l} p_k^{j-l} (p - p_k)^{l-v} + \sum_{v=1}^{n_k} \sum_{l=v}^{j} c_{kv} \binom{j}{l} p_k^{j-l} (p - p_k)^{l-v}\,. \tag{4.23}$$

Die zweite Summe in (4.22) bzw. (4.23) tritt nur auf, wenn $q \leqq j$ bzw. $v \leqq j$ ist. Nun ist wegen (4.22) für $t > 0$

$$\mathcal{L}^{-1}\left\{\sum_{a=1}^{m_k}\sum_{q=1}^{m_{ka}} d_{kaq}^{\varepsilon}\frac{p^j}{(p-p_{ka}(\varepsilon))^q}\right\} = \sum_{a=1}^{m_k}\sum_{q=1}^{m_{ka}}\sum_{l=0}^{q-1} d_{kaq}^{\varepsilon}\binom{j}{l} p_{ka}(\varepsilon)^{j-l}\frac{1}{(q-l-1)!}\,t^{q-l-1}\,e^{p_{ka}(\varepsilon)t}$$

und wegen (4.23) für $t > 0$

$$\mathcal{L}^{-1}\left\{\sum_{v=1}^{n_k} c_{kv}\frac{p^j}{(p-p_k)^v}\right\} = \sum_{v=1}^{n_k}\sum_{l=0}^{v-1} c_{kv}\binom{j}{l} p_k^{j-l}\frac{1}{(v-l-1)!}\,t^{v-l-1}\,e^{p_k t}.$$

Zum Beweis von (4.21) bleibt also zu zeigen, daß für $t > 0$

$$\lim_{\varepsilon\to+0}\sum_{a=1}^{m_k}\sum_{q=1}^{m_{ka}} d_{kaq}^{\varepsilon}\binom{j}{l} p_{ka}(\varepsilon)^{j-l}\frac{1}{(q-l-1)!}\,t^{q-l-1}\,e^{p_{ka}(\varepsilon)t} = \sum_{v=1}^{n_k}\sum_{l=0}^{v-1} c_{kv}\binom{j}{l} p_k^{j-l}\frac{1}{(v-l-1)!}\,t^{v-l-1}\,e^{p_k t} \tag{4.24}$$

gilt. Beide Seiten von (4.24) kann man so umformen, daß (4.24) die folgende Gestalt annimmt:

$$\lim_{\varepsilon\to+0}\sum_{a=1}^{m_k}\left[\frac{1}{(m_{ka}-1)!}\frac{(p-p_{ka}(\varepsilon))^{m_{ka}}}{\Delta_\varepsilon(p)}e^{pt}p^j\right]^{(m_{ka}-1)}_{p=p_{ka}(\varepsilon)} = \frac{1}{(n_k-1)!}\left[\frac{(p-p_k)^{n_k}}{\Delta(p)}e^{pt}p^j\right]^{(n_k-1)}_{p=p_k}. \tag{4.25}$$

Aus Hilfssatz 4.2 folgt die Gültigkeit von (4.25). Damit ist der Satz 4.1 bewiesen.

5. Approximationssatz für das Problem $[A_{\overline{II}}]$

Zunächst werden einige Formeln mit Binomialkoeffizienten bewiesen, die beim Beweis des Approximationssatzes benötigt werden.

1.
$$\sum_{s=0}^{j}\binom{j}{s}\binom{s-q}{g}(-1)^s = 0,\qquad g = 0,\dots,j-1,\qquad j\geqq 1, \tag{5.1}$$

$$\sum_{s=0}^{j}\binom{j}{s}\binom{s-q}{j}(-1)^s = (-1)^j. \tag{5.2}$$

Beweis: Wir betrachten die Funktion

$$\Phi(t) = \sum_{s=0}^{j}\binom{j}{s}(-t)^s\ (j\geqq 1),$$

dann ist

$$\Phi^{(r)}(t) = j\cdot(j-1)\cdot\ldots\cdot(j-(r-1))\,(-1)^r\sum_{l=0}^{j-r}\binom{j-r}{l}(-t)^l,$$

also

$$\Phi^{(r)}(1) = 0\quad\text{für}\quad 0\leqq r\leqq j-1,\qquad \Phi^{(j)}(1) = j!(-1)^j. \tag{5.3}$$

Wir setzen

$$B_g = \sum_{s=0}^{j}\binom{j}{s}\binom{s-q}{g}(-1)^s,\qquad B(t) = \sum_{s=0}^{j}\binom{j}{s}(-t)^{s-q},$$

dann ist

$$B^{(g)}(1) = (-1)^{-q}g!\,B_g$$

und

$$B(t)\,(-t)^q = \Phi(t).$$

Hieraus folgt wegen (5.3) $B_g = 0$ für $g = 0,\dots,j-1$ und $B_j = (-1)^j$.

2.
$$\sum_{s=0}^{j}\binom{j}{s}\binom{g+q-s-1}{g}(-1)^s = 0,\qquad g = 0,\dots,j-1,\qquad j\geqq 1, \tag{5.4}$$

$$\sum_{s=0}^{j}\binom{j}{s}\binom{j+q-s-1}{j}(-1)^s = 1. \tag{5.5}$$

Beweis: Man ersetze in (5.1) s durch $j-s$ und q durch $j-g-q+1$, dann folgt (5.4). Ersetzt man in (5.2) s durch $j-s$ und q durch $-q+1$, dann folgt (5.5).

3.
$$\sum_{s=0}^{q-1}\binom{j}{s}\binom{g+q-s-1}{g}(-1)^s = 0,\qquad j\geqq q>1,\qquad j-q+1\leqq g<j, \tag{5.6}$$

$$\sum_{s=0}^{q-1}\binom{j}{s}\binom{j+q-s-1}{j}(-1)^s = 1,\qquad j\geqq q\geqq 1. \tag{5.7}$$

Beweis: In (5.4) und (5.5) sind unter den gemachten Voraussetzungen die zweiten Binomialkoeffizienten Null für $s = q, \dots, s = j$.

4. $$\sum_{s=0}^{q-1} \binom{j}{s} \binom{g+q-s-1}{g} (-1)^s + \sum_{s=q+g}^{j} \binom{j}{s} \binom{s-q}{g} (-1)^{s-g} = 0, \qquad j \geqq q \geqq 1, \qquad 0 \leqq g \leqq j-q. \tag{5.8}$$

Beweis: Wir setzen

$$C_g = \sum_{s=0}^{q-1} \binom{j}{s} \binom{g+q-s-1}{g} (-1)^s + \sum_{s=q+g}^{j} \binom{j}{s} \binom{s-q}{g} (-1)^{s-g}$$

und

$$C(t) = \sum_{s=0}^{q-1} \binom{j}{s} \frac{1}{(-t)^{q-s}} + \sum_{s=q}^{j} \binom{j}{s} (-t)^{s-q}.$$

Dann ist

$$C^{(g)}(1) = \frac{1}{(-1)^q} (-1)^g g!\, C_g$$

und

$$C(t)\,(-t)^q = \Phi(t).$$

Mit (5.3) folgt die Behauptung.

Neben dem Problem $[A_{\bar{II}}]$ betrachten wir das gestörte Problem A_{II}^{ε}:

Differentialgleichungssystem (vgl. (4.1)):

$$L_\varepsilon y_\varepsilon = k(t),$$

Anfangsbedingungen:

$$y_\varepsilon(-0) = 0, \dots, y_\varepsilon^{(n-1)}(-0) = 0.$$

Der Operator L_ε erfülle dieselben Voraussetzungen wie im Abschnitt 4. Wir müssen die Fälle A) Grad $\Delta(p) = n_0 \geqq 1$ und B) $\Delta(p) = p_0 = \text{const}$ unterscheiden. Im Fall A verwenden wir für die Nullstellen von $\Delta(p)$ und die Nullstellen, in die diese durch die ε-Störung aufspalten, dieselben Bezeichnungen wie in 4. Bei den restlichen Nullstellen müssen wir jetzt die Vielfachheiten genau berücksichtigen. Die restlichen verschiedenen Nullstellen von $\Delta_\varepsilon(p)$ bezeichnen wir mit $p_{n_0+1}(\varepsilon), \dots, p_{n_0+n^*}(\varepsilon)$, ihre Vielfachheiten mit $m_{n_0+1}, \dots, m_{n_0+n^*} (m_{n_0+1} + \dots + m_{n_0+n^*} = n \cdot m - n_0)$.

Im Fall B bezeichnen wir die verschiedenen Nullstellen von $\Delta_\varepsilon(p)$ mit $p_1(\varepsilon), \dots, p_{n^*}(\varepsilon)$ und ihre Vielfachheiten mit $m_1, \dots, m_{n^*} (m_1 + \dots + m_{n^*} = n \cdot m)$.

Satz 5.1: *Es erfülle $k(t)$ die Voraussetzungen des Satzes 2.6 und es sei darüber hinaus $k^{(u)}(t)$, $u = (m-1)\,n - 1$ für $t > 0$ stetig und $\mathcal{L}$-transformierbar, schließlich sei $y \in \mathcal{D}'_{m+}$ die Lösung des Problems $[A_{\bar{II}}]$ und $y_\varepsilon(t)$ die Lösung des gestörten Problems A_{II}^{ε}.*

A) *Im Falle* Grad $\Delta(p) = n_0 \geqq 1$ *sei noch $k^{(r)}(t)$, $r = (m-1)\,n + \max\limits_{a=1,\dots,n^*} m_{n_0+a}$, für $t > 0$ stetig, lokal integrabel und*

$$\lim_{\varepsilon \to +0} \operatorname{Re} p_{n_0+j}(\varepsilon) = -\infty, \qquad j = 1, \dots, n^*, \tag{5.9}$$

sowie

$$\lim_{\varepsilon \to +0} \frac{[p_{n_0+j}(\varepsilon)]^{q-1}}{\left[\dfrac{(p - p_{n_0+j}(\varepsilon))^{m_{n_0+j}}}{\Delta_\varepsilon(p)}\right]^{(m_{n_0+j}-q)}_{p=p_{n_0+j}(\varepsilon)}} \neq 0, \qquad j = 1, \dots, n^*, \qquad q = 1, \dots, m_{n_0+j}, \tag{5.10}$$

B) *Im Falle $\Delta(p) = p_0 = $ const sei noch $k^{(p)}(t)$, $p = (m-1)\,n + \max\limits_{a=1,\dots,n^*} m_a$, für $t > 0$ stetig, lokal integrabel und*

$$\lim_{\varepsilon \to +0} \operatorname{Re} p_j(\varepsilon) = -\infty, \qquad j = 1, \dots, n^*, \tag{5.11}$$

sowie

$$\lim_{\varepsilon \to +0} \frac{[p_j(\varepsilon)]^q}{\left[\dfrac{(p - p_j(\varepsilon))^{m_j}}{\Delta_\varepsilon(p)}\right]^{(m_j-q)}_{p=p_j(\varepsilon)}} \neq 0, \qquad j = 1, \dots, n^*, \qquad q = 1, \dots, m_j. \tag{5.12}$$

Dann ist

$$\lim_{\varepsilon \to +0} y_\varepsilon(t) = y(t) \quad \textit{für alle} \quad t > 0.$$

Bemerkung: Es ist in (5.10) und (5.12) nur verlangt, daß der Limes nicht Null ist, er darf also unendlich sein.

Beweis:

A) Mit den Bezeichnungen des Beweises von Satz 4.1 erhalten wir nach (2.16) $\left(K_l(p) = \mathscr{L}\{k_l(t)\}\right)$

$$\mathscr{L}\{y_s\} = Y_s(p) = \frac{1}{\Delta(p)} \sum_{l=1}^{m} A_{ls}(p)\, K_l(p) \tag{5.13}$$

und

$$\mathscr{L}\{y_s^\varepsilon\} = Y_s^\varepsilon(p) =$$

$$= S_\varepsilon(p) \sum_{l=1}^{m} A_{ls}(p)\, K_l(p) + \varepsilon S_\varepsilon(p) \sum_{l=1}^{m} \hat{A}_{ls}^\varepsilon(p)\, K_l(p) + Q_\varepsilon(p) \sum_{l=1}^{m} A_{ls}(p)\, K_l(p) + \varepsilon Q_\varepsilon(p) \sum_{l=1}^{m} \hat{A}_{ls}^\varepsilon(p)\, K_l(p)\,. \tag{5.14}$$

Wir zeigen für $t > 0$ $\left(K(p)\right.$ ist eine der Komponenten $\left.K_l(p)\right)$

$$\lim_{\varepsilon\to+0} \mathscr{L}^{-1}\{S_\varepsilon(p)\, p^j K(p)\} = \mathscr{L}^{-1}\left\{\frac{1}{\Delta(p)}\, p^j K(p)\right\}, \qquad j = 0, \dots, (m-1)\, n\,, \tag{5.15}$$

dann gilt für $t > 0$

$$\lim_{\varepsilon\to+0} \mathscr{L}^{-1}\left\{S_\varepsilon(p) \sum_{l=1}^{m} A_{ls}(p)\, K_l(p)\right\} = \mathscr{L}^{-1}\left\{\frac{1}{\Delta(p)} \sum_{l=1}^{m} A_{ls}(p)\, K_l(p)\right\}$$

und

$$\lim_{\varepsilon\to+0} \mathscr{L}^{-1}\left\{\varepsilon S_\varepsilon(p) \sum_{l=1}^{m} \hat{A}_{ls}^\varepsilon(p)\, K_l(p)\right\} = 0\,.$$

Zum Beweis von (5.15) genügt zu zeigen

$$\lim_{\varepsilon\to+0} \mathscr{L}^{-1}\left\{\sum_{a=1}^{m_k} \sum_{q=1}^{m_{ka}} d_{kaq}^\varepsilon \frac{p^j}{(p - p_{ka}(\varepsilon))^q} K(p)\right\} = \mathscr{L}^{-1}\left\{\sum_{v=1}^{n_k} c_{kv} \frac{p^j}{(p-p_k)^v} K(p)\right\}. \tag{5.16}$$

Nach (4.22) ist unter Verwendung des binomischen Lehrsatzes

$$\sum_{a=1}^{m_k} \sum_{q=1}^{m_{ka}} d_{kaq}^\varepsilon \frac{p^j}{(p - p_{ka}(\varepsilon))^q} K(p) = \sum_{a=1}^{m_k} \sum_{q=1}^{m_{ka}} \sum_{l=0}^{q-1} d_{kaq}^\varepsilon \binom{j}{l} p_{ka}(\varepsilon)^{j-l} (p - p_{ka}(\varepsilon))^{l-q} K(p) +$$

$$+ \sum_{a=1}^{m_k} \sum_{q=1}^{m_{ka}} \sum_{l=q}^{j} d_{kaq}^\varepsilon \binom{j}{l} p_{ka}(\varepsilon)^{j-l} \sum_{u=0}^{l-q} \binom{l-q}{u} (-1)^{l-q-u} p_{ka}(\varepsilon)^{l-q-u} p^u K(p)\,,$$

also erhält man aus dem Faltungssatz, (2.13) und Satz 2.1

$$\mathscr{L}^{-1}\left\{\sum_{a=1}^{m_k} \sum_{q=1}^{m_{ka}} d_{kaq}^\varepsilon \frac{p^j}{(p - p_{ka}(\varepsilon))^q} K(p)\right\}$$

$$= \sum_{a=1}^{m_k} \sum_{q=1}^{m_{ka}} \sum_{l=0}^{q-1} d_{kaq}^\varepsilon \binom{j}{l} p_{ka}(\varepsilon)^{j-l} \frac{1}{(q-l-1)!} t^{q-l-1} e^{p_{ka}(\varepsilon)t} * k(t) +$$

$$+ \sum_{a=1}^{m_k} \sum_{q=1}^{m_{ka}} \sum_{l=q}^{l} \sum_{u=0}^{l-q} d_{kaq}^\varepsilon \binom{j}{l} \binom{l-q}{u} (-1)^{l-q-u} p_{ka}(\varepsilon)^{j-q-u} k^{(u)}(t) +$$

$$+ \sum_{a=1}^{m_k} \sum_{q=1}^{m_{ka}} \sum_{l=q}^{j} \sum_{u=0}^{l-q} \sum_{g=0}^{u-1} d_{kaq}^\varepsilon \binom{j}{l} \binom{l-q}{u} (-1)^{l-q-u} p_{ka}(\varepsilon)^{j-q-u} k^{(g)}(+0)\, \delta^{(u-1-g)}(t)\,. \tag{5.17}$$

Analog erhält man aus (4.23)

$$\sum_{v=1}^{n_k} c_{kv} \frac{p^j}{(p-p_k)^v} K(p)$$

$$= \sum_{v=1}^{n_k} \sum_{l=0}^{v-1} c_{kv} \binom{j}{l} p_k^{j-l} (p-p_k)^{l-v} K(p) + \sum_{v=1}^{n_k} \sum_{l=v}^{j} c_{kv} \binom{j}{l} p_k^{j-l} \sum_{u=0}^{l-v} \binom{l-v}{u} (-1)^{l-v-u} p_k^{l-v-u} p^u K(p)$$

und

$$\mathscr{L}^{-1}\left\{\sum_{v=1}^{n_k} c_{kv} \frac{p^j}{(p-p_k)^v} K(p)\right\}$$

$$= \sum_{v=1}^{n_k} \sum_{l=0}^{v-1} c_{kv} \binom{j}{l} p_k^{j-l} \frac{1}{(v-l-1)!} t^{v-l-1} e^{p_k t} * k(t) +$$

$$+ \sum_{v=1}^{n_k} \sum_{l=v}^{j} \sum_{u=0}^{l-v} c_{kv} \binom{j}{l} \binom{l-v}{u} (-1)^{l-v-u} p_k^{j-v-u} k^{(u)}(t) +$$

$$+ \sum_{v=1}^{n_k} \sum_{l=v}^{j} \sum_{u=0}^{l-v} \sum_{g=0}^{u-1} c_{kv} \binom{j}{l} \binom{l-v}{u} (-1)^{l-v-u} p_k^{j-v-u} k^{(g)}(+0)\, \delta^{(u-1-g)}(t)\,. \tag{5.18}$$

(4.24) gilt offenbar gleichmäßig bezüglich $t \in [0, a]$, so daß daraus folgt

$$\lim_{\varepsilon\to+0} \sum_{a=1}^{m_k} \sum_{q=1}^{m_{ka}} \sum_{l=0}^{q-1} d_{kaq}^{\varepsilon} \binom{j}{l} p_{ka}(\varepsilon)^{j-l} \frac{1}{(q-l-1)!} t^{q-l-1} e^{p_{ka}(\varepsilon)t} * k(t)$$

$$= \sum_{v=1}^{n_k} \sum_{l=0}^{v-1} c_{kv} \binom{j}{l} p_k^{j-l} \frac{1}{(v-l-1)!} t^{v-l-1} e^{p_k t} * k(t) . \tag{5.19}$$

Nun ist noch der Grenzübergang des zweiten Ausdrucks in (5.17) zu untersuchen. Unter Beachtung von (5.1) erhält man, da $u = 0, \dots, j-1$ ist,

$$\sum_{a=1}^{m_k} \sum_{q=1}^{m_{ka}} \sum_{l=q}^{j} d_{kaq}^{\varepsilon} \binom{j}{l} \binom{l-q}{u} (-1)^{l-q-u} p_{ka}(\varepsilon)^{j-q-u} = -\sum_{a=1}^{m_k} \sum_{q=1}^{m_{ka}} \sum_{l=0}^{q-1} d_{kaq}^{\varepsilon} \binom{j}{l} \binom{l-q}{u} (-1)^{l-q-u} p_{ka}(\varepsilon)^{j-q-u}$$

$$= + \sum_{a=1}^{m_k} \frac{1}{(m_{ka}-1)!} \left[\frac{(p-p_{ka}(\varepsilon))^{m_{ka}}}{\Delta_\varepsilon(p)} p^{j-u-1} \right]_{p=p_{ka}(\varepsilon)}^{(m_{ka}-1)}$$

und analog

$$\sum_{v=1}^{n_k} \sum_{l=v}^{j} c_{kv} \binom{j}{l} \binom{l-v}{u} (-1)^{l-v-u} p_k^{j-v-u} = -\sum_{v=1}^{n_k} \sum_{l=0}^{v-1} c_{kv} \binom{j}{l} \binom{l-v}{u} (-1)^{l-v-u} p_k^{j-v-u}$$

$$= + \frac{1}{(n_k-1)!} \left[\frac{(p-p_k)^{n_k}}{\Delta(p)} p^{j-u-1} \right]_{p=p_k}^{(n_k-1)} .$$

Aus Hilfssatz 4.2 folgt nun, daß der zweite Ausdruck in (5.17) für $\varepsilon \to +0$ gegen den zweiten Ausdruck in (5.18) konvergiert. Damit ist (5.15) bewiesen.

Teil A des Satzes ist bewiesen, wenn wir noch zeigen

$$\lim_{\varepsilon\to+0} \mathcal{L}^{-1}\{Q_\varepsilon(p)\, p^j K(p)\} = 0 \quad \text{für} \quad t > 0 , \qquad j = 0, \dots, (m-1)\, n . \tag{5.20}$$

Nun ist

$$Q_\varepsilon(p) = \sum_{a=1}^{n^*} \sum_{q=1}^{m_{n_0+a}} d_{aq}^{\varepsilon} \frac{1}{(p-p_{n_0+a}(\varepsilon))^q} \tag{5.21}$$

mit

$$d_{aq}^{\varepsilon} = \frac{1}{(m_{n_0+a}-q)!} \left[\frac{(p-p_{n_0+a}(\varepsilon))^{m_{n_0+a}}}{\Delta_\varepsilon(p)} \right]_{p=p_{n_0+a}(\varepsilon)}^{(m_{n_0+a}-q)} .$$

Zum Beweis von (5.20) genügt zu zeigen

$$\lim_{\varepsilon\to+0} \mathcal{L}^{-1} \left\{ d_{aq}^{\varepsilon} \frac{p^j}{(p-p_{n_0+a}(\varepsilon))^q} K(p) \right\} = 0 , \qquad t > 0 , \qquad a = 1, \dots, n^* , \qquad q = 1, \dots, m_{n_0+a} . \tag{5.22}$$

Nun ist

$$d_{aq}^{\varepsilon} \frac{p^j}{(p-p_{n_0+a}(\varepsilon))^q} K(p) = d_{aq}^{\varepsilon} \sum_{s=0}^{q-1} \binom{j}{s} p_{n_0+a}(\varepsilon)^{j-s} \frac{1}{(p-p_{n_0+a}(\varepsilon))^{q-s}} K(p) +$$

$$+ d_{aq}^{\varepsilon} \sum_{s=q}^{j} \binom{j}{s} p_{n_0+a}(\varepsilon)^{j-s} \sum_{u=0}^{s-q} \binom{s-q}{u} (-1)^{s-q-u} p_{n_0+a}(\varepsilon)^{s-q-u} p^u K(p) ,$$

also erhält man für $t > 0$ (vgl. (5.17))

$$\mathcal{L}^{-1} \left\{ d_{aq}^{\varepsilon} \frac{p^j}{(p-p_{n_0+a}(\varepsilon))^q} K(p) \right\} = d_{aq}^{\varepsilon} \sum_{s=0}^{q-1} \binom{j}{s} p_{n_0+a}(\varepsilon)^{j-s} \frac{1}{(q-s-1)!} t^{q-s-1} e^{p_{n_0+a}(\varepsilon)t} * k(t) +$$

$$+ d_{aq}^{\varepsilon} \sum_{s=q}^{j} \binom{j}{s} p_{n_0+a}(\varepsilon)^{j-s} \sum_{u=0}^{s-q} \binom{s-q}{u} (-1)^{s-q-u} p_{n_0+a}(\varepsilon)^{s-q-u} k^{(u)}(t) =: F_\varepsilon .$$

Durch $(q+j-s)$ malige partielle Integration der Faltung und Vertauschung der Summationsreihenfolge in der zweiten Summe folgt

$$F_\varepsilon = d_{aq}^{\varepsilon} \sum_{s=0}^{q-1} \binom{j}{s} p_{n_0+a}(\varepsilon)^{j-s} \frac{1}{(q-s-1)!} \sum_{l=1}^{q+j-s} (-1)^{l-1} \frac{1}{p_{n_0+a}(\varepsilon)^l} \left[e^{p_{n_0+a}(\varepsilon)\tau} \frac{d}{d\tau^{l-1}} \left(\tau^{q-s-1} k(t-\tau)\right) \right]_0^t +$$

$$+ d_{aq}^{\varepsilon} \sum_{s=0}^{q-1} \binom{j}{s} p_{n_0+a}(\varepsilon)^{j-s} \frac{1}{(q-s-1)!} (-1)^{q+j-s} \frac{1}{p_{n_0+a}(\varepsilon)^{q+j-s}} \int_0^t e^{p_{n_0+a}(\varepsilon)\tau} \frac{d}{d\tau^{q+j-s}} \left(\tau^{q-s-1} k(t-\tau)\right) d\tau +$$

$$+ d_{aq}^{\varepsilon} \sum_{u=0}^{j-q} \sum_{s=q+u}^{j} \binom{j}{s} \binom{s-q}{u} (-1)^{s-q-u} p_{n_0+a}(\varepsilon)^{j-q-u} k^{(u)}(t) .$$

Auf Grund der Voraussetzungen (5.9) und (5.10) gilt

$$\lim_{\varepsilon\to+0} d_{aq}^{\varepsilon} \sum_{s=0}^{q-1} \binom{j}{s} p_{n_0+a}(\varepsilon)^{-q} \frac{1}{(q-s-1)!} (-1)^{q+j-s} \int_0^t e^{p_{n_0+a}(\varepsilon)\tau} \frac{d}{d\tau^{q+j-s}} \left(\tau^{q-s-1} k(t-\tau)\right) d\tau = 0\,. \tag{5.23}$$

Nun ist

$$\left[e^{p_{n_0+a}(\varepsilon)\tau} \frac{d}{d\tau^{l-1}} \left(\tau^{q-s-1}k(t-\tau)\right)\right]_0^t = \left[e^{p_{n_0+a}(\varepsilon)\tau} \sum_{v=0}^{l-1} \binom{l-1}{v} (-1)^v k^{(v)}(t-\tau) \frac{d}{d\tau^{l-1-v}} \tau^{q-s-1}\right]_0^t$$

$$= e^{p_{n_0+a}(\varepsilon)t} \sum_{v=0}^{l-1} \binom{l-1}{v} (-1)^v k^{(v)}(+0) \frac{d}{dt^{l-1-v}} t^{q-s-1} - \binom{l-1}{l-q+s} (-1)^{l-q+s} k^{(l-q+s)}(t)\,(q-s-1)!\,.$$

Der zweite Ausdruck kann nur auftreten für $l \geqq q - s$.

Wegen der Voraussetzung (5.9) ist für $t > 0$

$$\lim_{\varepsilon\to+0} d_{aq}^{\varepsilon} \sum_{s=0}^{q-1} \binom{j}{s} p_{n_0+a}(\varepsilon)^{j-s} \frac{1}{(q-s-1)!} \sum_{l=1}^{q+j-s} (-1)^{l-1} \frac{1}{p_{n_0+a}(\varepsilon)^e} \times$$

$$\times\, e^{p_{n_0+a}(\varepsilon)t} \sum_{v=0}^{l-1} \binom{l-1}{v} (-1)^v k^{[v]}(+0) \frac{d}{dt^{l-1-v}} t^{q-s-1} = 0\,. \tag{5.24}$$

Also bleibt

$$F_\varepsilon = d_{aq}^{\varepsilon} \sum_{s=0}^{q-1} \binom{j}{s} p_{n_0+a}(\varepsilon)^{j-s} \sum_{l=q-s}^{q+j-s} \frac{(-1)^{s-q}}{p_{n_0+a}(\varepsilon)^l} \binom{l-1}{l-q+s} k^{(l-q+s)}(t) +$$

$$+ d_{aq}^{\varepsilon} \sum_{u=0}^{j-q} \sum_{s=q+u}^{j} \binom{j}{s} \binom{s-q}{u} (-1)^{s-q-u} p_{n_0+a}(\varepsilon)^{j-q-u} k^{(u)}(t) + o(1) =$$

$$= d_{aq}^{\varepsilon} \sum_{g=0}^{j} \sum_{s=0}^{q-1} \binom{j}{s} \binom{g+q-s-1}{g} (-1)^{s-q} p_{n_0+a}(\varepsilon)^{j-g-q} k^{(g)}(t) +$$

$$+ d_{aq}^{\varepsilon} \sum_{g=0}^{j-q} \sum_{s=q+g}^{j} \binom{j}{s} \binom{s-q}{g} (-1)^{s-q-g} p_{n_0+a}(\varepsilon)^{j-g-q} k^{(g)}(t) + o(1)\,. \tag{5.25}$$

Es sei $j < q$. Dann ist

$$F_\varepsilon = d_{aq}^{\varepsilon} \sum_{g=0}^{j} \sum_{s=0}^{j} \binom{j}{s} \binom{g+q-s-1}{g} (-1)^{s-q} p_{n_0+a}(\varepsilon)^{j-g-q} k^{(g)}(t) + o(1)\,.$$

Für $j = 0$ ist wegen (5.9) und (5.10) $\lim\limits_{\varepsilon\to+0} F_\varepsilon = 0$.

Für $j \geqq 1$ ist wegen (5.4) und (5.5)

$$F_\varepsilon = d_{aq}^{\varepsilon}(-1)^{-q} p_{n_0+a}(\varepsilon)^{-q} k^{(j)}(t) + o(1)\,,$$

also ist wegen (5.9) und (5.10) $\lim\limits_{\varepsilon\to+0} F_\varepsilon = 0$.

Es sei nun $j \geqq q \geqq 1$. Dann ist

$$F_\varepsilon = d_{aq}^{\varepsilon} \sum_{g=j-q+1}^{j} \sum_{s=0}^{q-1} \binom{j}{s} \binom{g+q-s-1}{g} (-1)^{s-q} p_{n_0+a}(\varepsilon)^{j-g-q} k^{(g)}(t) +$$

$$+ d_{aq}^{\varepsilon} \sum_{g=0}^{j-q} p_{n_0+a}(\varepsilon)^{j-g-q} k^{(g)}(t) \left[\sum_{s=0}^{q-1} \binom{j}{s} \binom{g+q-s-1}{g} (-1)^{s-q} + \sum_{s=q+g}^{j} \binom{j}{s} \binom{s-q}{g} (-1)^{s-q-g}\right] + o(1)$$

$$= d_{aq}^{\varepsilon} \sum_{g=j-q+1}^{j} \sum_{s=0}^{q-1} \binom{j}{s} \binom{g+q-s-1}{g} (-1)^{s-q} p_{n_0+a}(\varepsilon)^{j-g-q} k^{(g)}(t) + o(1) \tag{5.26}$$

wegen (5.8). Für $q = 1$ ist nach (5.26) $F_\varepsilon = -d_{a1}^{\varepsilon} p_{n_0+a}(\varepsilon)^{-1} k^{(j)}(t) + o(1)$, also ist auf Grund der Voraussetzungen (5.9), (5.10) $\lim\limits_{\varepsilon\to+0} F_\varepsilon = 0$. Für $j \geqq q > 1$ ist wegen (5.6) und (5.7) $F_\varepsilon = d_{aq}^{\varepsilon}(-1)^{-q} p_{n_0+a}(\varepsilon)^{-q} k^{(j)}(t) + o(1)$, also ist auf Grund von (5.9), (5.10) $\lim\limits_{\varepsilon\to+0} F_\varepsilon = 0$. Damit ist der Satz im Fall A bewiesen.

B) Es ist

$$\mathcal{L}\{y_s\} = Y_s(p) = \frac{1}{p_0} \sum_{l=1}^{m} A_{ls}(p)\, K_l(p)$$

und

$$\mathcal{L}\{y_s^{\varepsilon}\} = Y_s^{\varepsilon}(p) = \frac{1}{\Delta_\varepsilon(p)} \sum_{l=1}^{m} A_{ls}(p)\, K_l(p) + \varepsilon \frac{1}{\Delta_\varepsilon(p)} \sum_{l=1}^{m} \hat{A}_{ls}^{\varepsilon}(p)\, K_l(p)\,.$$

Der Satz ist bewiesen, wenn wir zeigen, daß für $t > 0$ gilt

$$\lim_{\varepsilon\to+0} \mathcal{L}^{-1}\left\{\frac{1}{\Delta_\varepsilon(p)}\, p^j K(p)\right\} = \mathcal{L}^{-1}\left\{\frac{1}{p_0}\, p^j K(p)\right\} = \frac{1}{p_0}\, k^{(j)}(t)\,, \qquad j = 0, \dots, (m-1)\,n\,. \tag{5.27}$$

Nun ist

$$\frac{1}{\Delta_\varepsilon(p)} = \sum_{a=1}^{n^*} \sum_{q=1}^{m_a} \frac{d_{aq}^\varepsilon}{(p - p_a(\varepsilon))^q} \tag{5.28}$$

mit

$$d_{aq}^\varepsilon = \frac{1}{(m_a - q)!}\left[\frac{(p - p_a(\varepsilon))^{m_a}}{\Delta_\varepsilon(p)}\right]^{(m_a - q)}_{p = p_a(\varepsilon)}.$$

Dieselben Rechnungen wie oben von (5.22) bis (5.25) ergeben für $t > 0$

$$\mathcal{L}^{-1}\left\{\frac{1}{\Delta_\varepsilon(p)}\, p^j K(p)\right\} = \sum_{a=1}^{n^*} \sum_{q=1}^{m_a} d_{aq}^\varepsilon \sum_{g=0}^{j} \sum_{s=0}^{q-1} \binom{j}{s}\binom{g+q-s-1}{g} (-1)^{s-q}\, p_a(\varepsilon)^{j-g-q}\, k^{(g)}(t) +$$
$$+ \sum_{a=1}^{n^*} \sum_{q=1}^{m_a} d_{aq}^\varepsilon \sum_{g=0}^{j-q} \sum_{s=q+g}^{j} \binom{j}{s}\binom{s-q}{g} (-1)^{s-q-g}\, p_a(\varepsilon)^{j-g-q}\, k^{(g)}(t) + o(1) =: \sum_{a=1}^{n^*} \sum_{q=1}^{m_a} F_{aq}^\varepsilon + o(1)\,. \tag{5.29}$$

Hierbei ist zu beachten, daß (5.23) auch unter den Voraussetzungen (5.11) und (5.12) für $\varepsilon \to +0$ gegen Null geht, denn der Faktor vor dem Integral bleibt wegen (5.12) beschränkt, und das Integral geht wegen (5.11) gegen Null.

Für $j = 0$ erhalten wir aus (5.29)

$$F_{aq}^\varepsilon = d_{aq}^\varepsilon (-1)^q\, p_a(\varepsilon)^{-q}\, k^{(j)}(t)\,. \tag{5.30}$$

Für $1 \leqq j < q$ ist wegen (5.4) und (5.5)

$$F_{aq}^\varepsilon = d_{aq}^\varepsilon \sum_{g=0}^{j} \sum_{s=0}^{j} \binom{j}{s}\binom{g+q-s-1}{g} (-1)^{s-q}\, p_a(\varepsilon)^{j-g-q}\, k^{(g)}(t)$$
$$= d_{aq}^\varepsilon \sum_{s=0}^{j} \binom{j}{s}\binom{j+q-s-1}{j} (-1)^{s-q}\, p_a(\varepsilon)^{-q}\, k^{(j)}(t) = d_{aq}^\varepsilon (-1)^q\, p_a(\varepsilon)^{-q}\, k^{(j)}(t)\,. \tag{5.31}$$

Für $j \geqq q \geqq 1$ ist wegen (5.8) (vgl. (5.26))

$$F_{aq}^\varepsilon = d_{aq}^\varepsilon \sum_{g=j-q+1}^{j} \sum_{s=0}^{q-1} \binom{j}{s}\binom{g+q-s-1}{g} (-1)^{s-q}\, p_a(\varepsilon)^{j-g-q}\, k^{(g)}(t)\,. \tag{5.32}$$

Für $j \geqq q = 1$ folgt aus (5.32)

$$F_{aq}^\varepsilon = d_{aq}^\varepsilon (-1)^q\, p_a(\varepsilon)^{-q}\, k^{(j)}(t)\,,$$

und für $j \geqq q > 1$ ist wegen (5.6) und (5.7)

$$F_{aq}^\varepsilon = d_{aq}^\varepsilon (-1)^q\, p_a(\varepsilon)^{-q}\, k^{(j)}(t)\,. \tag{5.33}$$

Damit ist für $t > 0$ (beachte (5.28))

$$\mathcal{L}^{-1}\left\{\frac{1}{\Delta_\varepsilon(p)}\, p^j K(p)\right\} = \sum_{a=1}^{n^*} \sum_{q=1}^{m_a} d_{aq}^\varepsilon (-1)^q\, p_a(\varepsilon)^{-q}\, k^{(j)}(t) + o(1) = \left.\frac{1}{\Delta_\varepsilon(p)}\right|_{p=0} k^{(j)}(t) + o(1)\,,$$

also gilt für $t > 0$

$$\lim_{\varepsilon\to+0} \mathcal{L}^{-1}\left\{\frac{1}{\Delta_\varepsilon(p)}\, p^j K(p)\right\} = \frac{1}{p_0}\, k^{(j)}(t)\,.$$

Damit ist der Satz 5.1 vollständig bewiesen.

6. Anormales System mit δ-Distributionen

Wir beschäftigen uns jetzt noch einmal mit dem Problem $[A_\delta^-]$ im anormalen Fall und der Approximation der Lösung des Problems $[A_\delta^-]$ durch die eines stabil gestörten Systems. Wie in 4. setzen wir voraus $\det A^{(n)} = 0$, $\Delta(p) \not\equiv 0$, und betrachten daneben ein *gestörtes* Problem $[A_\delta^\varepsilon]$:

$$A_\varepsilon^{(n)} D^n y + \dots + A_\varepsilon^{(1)} D y + A_\varepsilon^{(0)} y = \sum_{i=0}^{N} \alpha_i [\delta^{(i)}]\,, \qquad y \in \mathcal{D}'_{m+}\,.$$

Wir verwenden dieselben Bezeichnungen wie in 4. Nach Satz 2.6 hat das Problem $[A_\delta^-]$ eine eindeutige Lösung $y \in \mathcal{D}'_{m+}$ und nach Satz 2.5 hat das Problem $[A_\delta^\varepsilon]$ eine eindeutige Lösung $y_\varepsilon \in \mathcal{D}'_{m+}$.

Satz 6.1: *Sei $y \in \mathcal{D}'_{m+}$ die Lösung des Problems* $[\mathrm{A}_\delta^-]$ *und* $y_\varepsilon \in \mathcal{D}'_{m+}$ *die Lösung des Problems* $[\mathrm{A}_\delta^\varepsilon]$. *Falls*

$$\lim_{\varepsilon\to+0} \operatorname{Re} p_j(\varepsilon) = -\infty\,, \qquad j = n_0 + 1, \ldots, n\cdot m\,, \tag{6.1}$$

dann gilt

$$\lim_{\varepsilon\to+0} y_\varepsilon(t) = y(t) \quad \textit{für alle} \quad t > 0\,.$$

Beweis: Es ist nach (2.16)

$$\mathcal{L}\{y_s\} = Y_s(p) = \frac{1}{\Delta(p)} \sum_{l=1}^{m} \sum_{i=0}^{N} A_{ls}(p)\, \alpha_{li} p^i$$

und (vgl. (4.17))

$$\mathcal{L}\{y_s^\varepsilon\} = Y_s^\varepsilon(p) = S_\varepsilon(p) \sum_{l=1}^{m} \sum_{i=0}^{N} A_{ls}(p)\, \alpha_{li} p^i + \varepsilon S_\varepsilon(p) \sum_{l=1}^{m} \sum_{i=0}^{N} \hat{A}_{ls}^\varepsilon(p)\, \alpha_{li} p^i +$$

$$+ Q_\varepsilon(p) \sum_{l=1}^{m} \sum_{i=0}^{N} A_{ls}(p)\, \alpha_{li} p^i + \varepsilon Q_\varepsilon(p) \sum_{l=1}^{m} \sum_{i=0}^{N} \hat{A}_{ls}^\varepsilon(p)\, \alpha_{li} p^i\,.$$

Der Beweis ergibt sich aus (4.19) und der Tatsache, daß wegen der Voraussetzung (6.1) Ausdrücke der Form (4.18) für $t > 0$ bei $\varepsilon \to +0$ gegen Null konvergieren.

Beispiel 6.1: Wir betrachten das zum Beispiel 3.1 stabil gestörte Problem

$$(y_1^\varepsilon)' + \varepsilon (y_2^\varepsilon)' + y_2^\varepsilon = 0\,,$$

$$-\varepsilon (y_1^\varepsilon)' + y_1^\varepsilon + y_2^\varepsilon = \delta'(t)\,.$$

Anfangsbedingungen: $y_1^\varepsilon(-0) = 0\,, \quad y_2^\varepsilon(-0) = 0\,.$

Es ist $\det A_\varepsilon^{(1)} = \varepsilon^2$, $\Delta_\varepsilon(p) = \varepsilon^2 p^2 + p - 1 = \varepsilon^2 (p - p_1(\varepsilon))\,(p - p_2(\varepsilon))$ mit

$$p_1(\varepsilon) = -\frac{1}{2\varepsilon^2} + \frac{1}{2\varepsilon^2}\sqrt{1 + 4\varepsilon^2} = 1 + O(\varepsilon^2)\,,$$

$$p_2(\varepsilon) = -\frac{1}{2\varepsilon^2} - \frac{1}{2\varepsilon^2}\sqrt{1 + 4\varepsilon^2} - \frac{1}{\varepsilon^2} = -1 + O(\varepsilon^2)\,.$$

Damit sind die Voraussetzungen des Satzes 6.1 erfüllt. Man erhält für $t \geqq 0$

$$y_1^\varepsilon(t) = -\frac{1}{\varepsilon}\,\delta(t) + \frac{1}{\sqrt{1+4\varepsilon^2}}\left[-\frac{1}{\varepsilon} + p_1(\varepsilon)\left(\frac{1}{\varepsilon} - 1\right)\right] e^{p_1(\varepsilon)t} + \frac{1}{\sqrt{1+4\varepsilon^2}}\left[\frac{1}{\varepsilon} - p_2(\varepsilon)\left(\frac{1}{\varepsilon} - 1\right)\right] e^{p_2(\varepsilon)t}\,,$$

$$y_2^\varepsilon(t) = \frac{1}{\varepsilon^2}\,\delta(t) + \frac{1}{\sqrt{1+4\varepsilon^2}}\, p_1(\varepsilon)^2\, e^{p_1(\varepsilon)t} - \frac{1}{\sqrt{1+4\varepsilon^2}}\, p_2(\varepsilon)^2\, e^{p_2(\varepsilon)t}\,.$$

Abbildung 2 zeigt $y_1^\varepsilon(t)$, $y_2^\varepsilon(t)$ für einige Werte von ε zusammen mit den Grenzkurven von Beispiel 3.1 für $t > 0$.

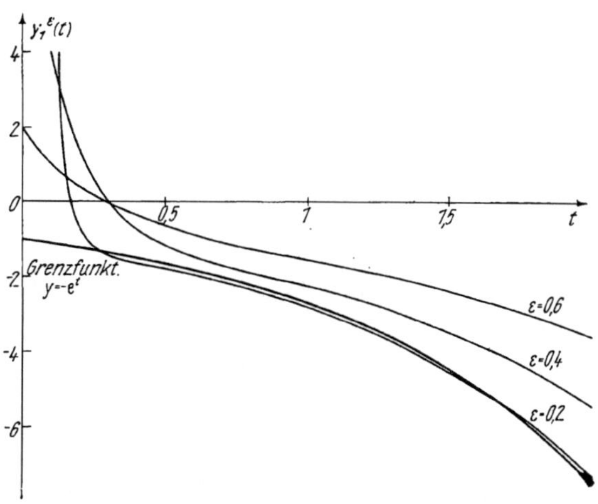

Abb. 2a

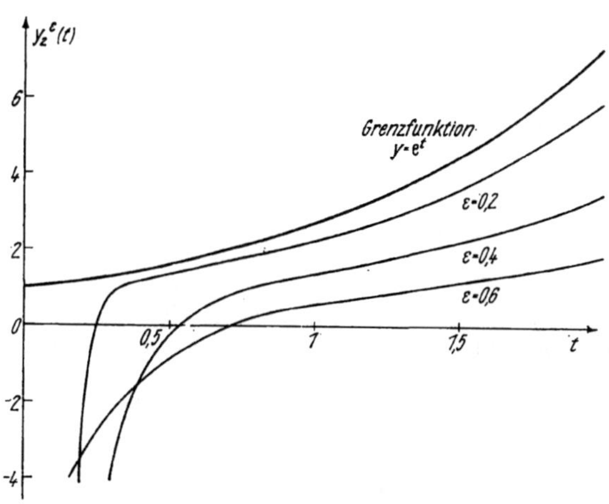

Abb. 2b

Wir approximieren nun noch wie in 3. $\sum_{i=0}^{N} \alpha_i[\delta^{(i)}]$ durch $k(t) = \sum_{i=0}^{N} \alpha_i h_\eta^{(i)}(t)$, wobei $k(t)$ die Voraussetzungen des Satzes 5.1 erfüllen soll. Mit $y_\eta \in \mathcal{D}'_{m+}$ bezeichnen wir die Lösung von (3.1) und mit $y_{\eta,\varepsilon} = y_{\eta,\varepsilon}(t)$ bezeichnen wir die klassische Lösung von $L_\varepsilon y_{\eta,\varepsilon} = \sum_{i=0}^{N} \alpha_i h_\eta^{(i)}(t)$ mit

$$y_{\eta,\varepsilon}(-0) = y_{\eta,\varepsilon}(+0) = 0\,, \dots, y_{\eta,\varepsilon}^{(n-1)}(-0) = y_{\eta,\varepsilon}^{(n-1)}(+0) = 0\,.$$

Dann gilt der

Satz 6.2: *Unter den Voraussetzungen des Satzes* 5.1 *gilt*

$$\lim_{\eta\to+0}\lim_{\varepsilon\to+0} y_{\eta,\varepsilon}(t) = \lim_{\varepsilon\to+0}\lim_{\eta\to+0} y_{\eta,\varepsilon}(t) = y(t) \quad \textit{für alle} \quad t > 0\,.$$

Beweis: Nach Satz 5.1 ist

$$\lim_{\varepsilon\to+0} y_{\eta,\varepsilon}(t) = y_\eta(t)\,,$$

und nach Satz 3.1 ist

$$\lim_{\eta\to+0} y_\eta(t) = y(t)\,.$$

Nach Satz 3.1 ist

$$\lim_{\eta\to+0} y_{\eta,\varepsilon}(t) = y_\varepsilon(t)\,,$$

und nach Satz 6.1 ist

$$\lim_{\varepsilon\to+0} y_\varepsilon(t) = y(t)\,.$$

Beispiel 6.2: Im Beispiel 3.1 nehmen wir jetzt eine stabile ε-Störung und eine Approximation von $\delta'(t)$ vor. Wir betrachten also das klassische Problem

$$(y_1^{\eta,\varepsilon})' + \varepsilon(y_2^{\eta,\varepsilon})' + y_2^{\eta,\varepsilon} = 0\,,$$

$$-\varepsilon(y_1^{\eta,\varepsilon})' + y_2^{\eta,\varepsilon} + y_2^{\eta,\varepsilon} = h_\eta'(t)\,, \qquad h_\eta(t) = \frac{30}{\eta^6}\,t(t-\eta)^4\,\big(s(t) - s(t-\eta)\big)\,,$$

mit den Anfangsbedingungen

$$y_1^{\eta,\varepsilon}(-0) = y_1^{\eta,\varepsilon}(+0) = 0\,, \qquad y_2^{\eta,\varepsilon}(-0) = y_2^{\eta,\varepsilon}(+0) = 0\,.$$

Die Bedingung (5.10) ist erfüllt, denn es ist (vgl. Beispiel 6.1)

$$\lim_{\varepsilon\to+0} \frac{p_2(\varepsilon)^0}{\left[\dfrac{p - p_2(\varepsilon)}{\Delta_\varepsilon(p)}\right]_{p=p_2(\varepsilon)}} = \lim_{\varepsilon\to+0} \varepsilon^2\big(p_2(\varepsilon) - p_1(\varepsilon)\big) = -1 \neq 0\,.$$

Also ist Satz 6.2 anwendbar.

Man erhält für $t \geqq 0$ (vgl. Beispiel 3.1)

$$y_1^{\eta,\varepsilon}(t) = \frac{30}{\eta^6}\left[K_1 + K_2 t + \frac{K_3}{2}t^2 + \frac{K_4}{6}t^3 + 5t^4 + K_5\,e^{p_1(\varepsilon)t} + K_6\,e^{p_2(\varepsilon)t}\right] -$$

$$-\frac{30}{\eta^6}\left[L_1 + L_2(t-\eta) + \frac{L_3}{2}(t-\eta)^2 + \frac{L_4}{6}(t-\eta)^3 + 5(t-\eta)^4 + L_5\,e^{p_1(\varepsilon)(t-\eta)} + L_6\,e^{p_2(\varepsilon)(t-\eta)}\right] s(t-\eta)$$

mit

$$K_1 = A + B\varepsilon + (C + D + 120)\,\varepsilon^2 + (D + 120)\,\varepsilon^3 + 120\varepsilon^4\,, \qquad K_2 = B + C\varepsilon + (D + 120)\,\varepsilon^2 + 120\varepsilon^3\,,$$

$$K_3 = C + D\varepsilon + 120\varepsilon^2\,, \qquad K_4 = D + 120\varepsilon\,,$$

$$K_5 = \frac{1}{\sqrt{1 + 4\varepsilon^2}}\,[p_2(\varepsilon)\,\varepsilon^2 K_1 - \varepsilon^2 K_2 - \eta^4\varepsilon]\,, \qquad K_6 = \frac{1}{\sqrt{1 + 4\varepsilon^2}}\,[-p_1(\varepsilon)\,\varepsilon^2 K_1 + \varepsilon^2 K_2 + \eta^4\varepsilon]\,,$$

$$L_1 = E + E\varepsilon + (2E + 120)\,\varepsilon^2 + (E + 120)\,\varepsilon^3 + 120\varepsilon^4\,, \qquad L_2 = E + E\varepsilon + (E + 120)\,\varepsilon^2 + 120\varepsilon^3\,,$$

$$L_3 = E + E\varepsilon + 120\varepsilon^2\,, \qquad L_4 = E + 120\varepsilon\,,$$

$$L_5 = \frac{1}{\sqrt{1 + 4\varepsilon^2}}\,[p_2(\varepsilon)\,\varepsilon^2 L_1 - \varepsilon^2 L_2]\,, \qquad L_6 = \frac{1}{\sqrt{1 + 4\varepsilon^2}}\,[-p_1(\varepsilon)\,\varepsilon^2 L_1 + \varepsilon^2 L_2]$$

und

$$y_2^{\eta,\varepsilon}(t) = -\frac{30}{\eta^6}\left[M_1 + M_2 t + \frac{M_3}{2} t^2 + 20t^3 + M_4\, e^{p_1(\varepsilon)t} + M_5\, e^{p_2(\varepsilon)t}\right] +$$

$$+\frac{30}{\eta^6}\left[N_1 + N_2(t-\eta) + \frac{N_3}{2}(t-\eta)^2 + 20(t-\eta)^3 + N_4\, e^{p_1(\varepsilon)(t-\eta)} + N_5\, e^{p_2(\varepsilon)(t-\eta)}\right] s(t-\eta)$$

mit

$$M_1 = B + (D + 120)\,\varepsilon^2\,, \qquad M_2 = C + 120\varepsilon^2\,, \qquad M_3 = D\,,$$

$$M_4 = \frac{1}{\sqrt{1+4\varepsilon^2}}\left(p_2(\varepsilon)\,\varepsilon^2 M_1 - \varepsilon^2 M_2 - \eta^4\right), \qquad M_5 = \frac{1}{\sqrt{1+4\varepsilon^2}}\left(-p_1(\varepsilon)\,\varepsilon^2 M_1 + \varepsilon^2 M_2 + \eta^4\right),$$

$$N_1 = E + (E + 120)\,\varepsilon^2\,, \qquad N_2 = E + 120\varepsilon^2\,, \qquad N_3 = E\,,$$

$$N_4 = \frac{1}{\sqrt{1+4\varepsilon^2}}\left(p_2(\varepsilon)\,\varepsilon^2 N_1 - \varepsilon^2 N_2\right), \qquad N_5 = \frac{1}{\sqrt{1+4\varepsilon^2}}\left(-p_1(\varepsilon)\,\varepsilon^2 N_1 + \varepsilon^2 N_2\right).$$

Man erkennt ohne Mühe, daß diese Lösung für $\varepsilon \to +0$, $t > 0$, gegen die Lösung $y_\eta(t)$ von Beispiel 3.1 konvergiert.

Für $t > \eta$ gilt

$$y_1^{\eta,\varepsilon}(t) = \frac{30}{\eta^6}\left(K_5 - L_5\, e^{-p_1(\varepsilon)\eta}\right) e^{p_1(\varepsilon)t} + \frac{30}{\eta^6}\left(K_6 - L_6\, e^{-p_2(\varepsilon)\eta}\right) e^{p_2(\varepsilon)t}$$

$$= \left\{\frac{1}{\sqrt{1+4\varepsilon^2}}\left[-\frac{1}{\varepsilon} + p_1(\varepsilon)\left(\frac{1}{\varepsilon} - 1\right)\right] + \frac{p_1(\varepsilon)^2\left(1 + p_1(\varepsilon)\,\varepsilon\right)}{\sqrt{1+4\varepsilon^2}}\,\eta - 30L_5 \sum_{n=1}^{\infty}\frac{(-p_1(\varepsilon)\,\eta)^n}{(n+6)!}\right\} e^{p_1(\varepsilon)t} +$$

$$+\left\{\frac{1}{\sqrt{1+4\varepsilon^2}}\left[\frac{1}{\varepsilon} - p_2(\varepsilon)\left(\frac{1}{\varepsilon} - 1\right)\right] - \frac{p_2(\varepsilon)^2\left(1 + p_2(\varepsilon)\,\varepsilon\right)}{\sqrt{1+4\varepsilon^2}}\,\eta - 30L_6 \sum_{n=1}^{\infty}\frac{(-p_2(\varepsilon)\,\eta)^n}{(n+6)!}\right\} e^{p_2(\varepsilon)t},$$

$$y_2^{\eta,\varepsilon}(t) = \frac{30}{\eta^6}\left(-M_4 + N_4\, e^{-p_1(\varepsilon)\eta}\right) e^{p_1(\varepsilon)t} + \frac{30}{\eta^6}\left(-M_5 + N_5\, e^{-p_2(\varepsilon)\eta}\right) e^{p_2(\varepsilon)t} =$$

$$= \left\{\frac{1}{\sqrt{1+4\varepsilon^2}}\left(p_1(\varepsilon)^2 - p_1(\varepsilon)^3\eta\right) + 30N_4 \sum_{n=1}^{\infty}\frac{(-p_1(\varepsilon)\,\eta)^n}{(n+6)!}\right\} e^{p_1(\varepsilon)t} +$$

$$+\left\{\frac{1}{\sqrt{1+4\varepsilon^2}}\left(-p_2(\varepsilon)^2 + p_2(\varepsilon)^3\eta\right) + 30N_5 \sum_{n=1}^{\infty}\frac{(-p_2(\varepsilon)\,\eta)^n}{(n+6)!}\right\} e^{p_2(\varepsilon)t}.$$

Man sieht hieran sofort, daß $y_{\eta,\varepsilon}(t)$ für $\eta \to +0$, $t > 0$, gegen die Lösung von Beispiel 6.1 konvergiert.

Abbildung 3 zeigt $y_1^{\eta,\varepsilon}(t)$, $y_2^{\eta,\varepsilon}(t)$ für einige Werte (η, ε) zusammen mit den Grenzkurven von Beispiel 3.1 für $t > 0$.

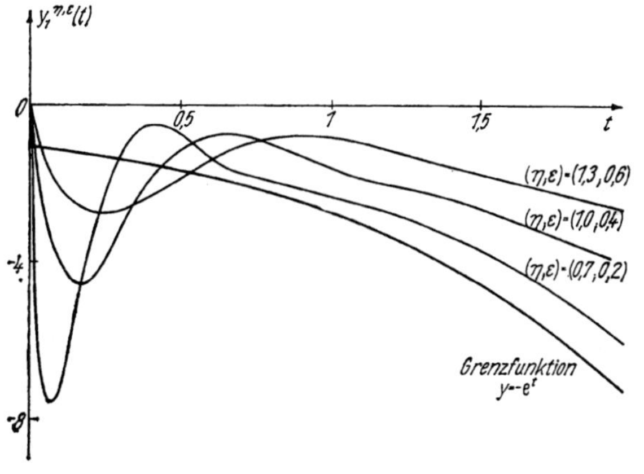

Abb. 3a

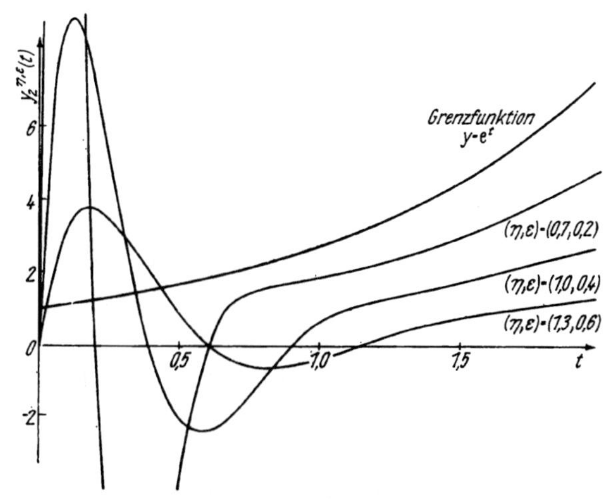

Abb. 3b

Literatur

1 MOHR, E., Integration von gewöhnlichen Differentialgleichungen mit konstanten Koeffizienten mittels Operatorenrechnung, Math. Nachr. 10 (1953), 1—49.
2 BERG, L., Operatorenrechnung I: Algebraische Methoden, VEB Deutscher Verlag der Wissenschaften, Berlin 1972.
3 DOETSCH, G., Handbuch der LAPLACE-Transformation. II. Teil, Birkhäuser Verlag Basel und Stuttgart 1955.
4 FOCKE, J., Lösung von Systemen gewöhnlicher Differentialgleichungen durch Sprungfunktionen, Wiss. Z. Karl-Marx-Univ. Leipzig 8 (1958/59), Math.-Naturw. Reihe, H. 5, 867—869.
5 KRABBE, G., Operational Calculus, A Plenum/Rosetta Edition 1975.
6 SAUER, R.; SZABÓ, I., Mathematische Hilfsmittel des Ingenieurs. Teil I, Springer-Verlag, Berlin-Heidelberg-New York 1967.
7 DOETSCH, G., Einführung in Theorie und Anwendung der LAPLACE-Transformation, Basel-Stuttgart, Birkhäuser 1958.
8 DOETSCH, G., Anleitung zum praktischen Gebrauch der LAPLACE-Transformation, R. Oldenbourg, München 1956.
9 BERG, L., Solution of degenerate linear initial value problems, ZAMM 57 (1977), 65—73.
10 SUPPE, C., Entartete Systeme linearer Differentialgleichungen erster Ordnung mit konstanten Koeffizienten als Grenzfall normaler Systeme, ZAMM 48 (1968), 286—288.
11 STOPP, F., Über die verallgemeinerten Lösungen bei entarteten Anfangswertproblemen für gewöhnliche Differentialgleichungen, ZAMM 48 (1968), 288—289.
12 BERESIN, I. S.; SHIDKOW, N. P., Numerische Methoden I, VEB Deutscher Verlag der Wissenschaften, Berlin 1970.

Frau Dipl.-Math. I. HEYNIG und Herrn Dipl.-Math. W. HAMBACH danke ich herzlich für das Durchrechnen der Beispiele und ihre graphische Auswertung. Leider konnten aus Platzgründen nicht alle vorbereiteten Beispiele aufgenommen werden.

Eingereicht: 4. 8. 1980, revidierte und gekürzte Fassung zweier am 24. 4. 1980 eingereichter Arbeiten.

Anschrift: Prof. Dr. sc. nat. L. JENTSCH, Technische Hochschule Karl-Marx-Stadt, Sektion Mathematik, Reichenhainer Str. 39—41, DDR 9010 Karl-Marx-Stadt

ZAMM 61, 161–164 (1981)

H. E. SALZER

Rational Interpolation Using Incomplete Barycentric Forms

Bei der rationalen Interpolation von $f(x)$ durch $(a_0 + a_1x + \dots + a_nx^n)/(1 + b_1x + \dots + b_mx^m)$ in x_i, $i = 1(1)\, n + m + 1$ lösen wir anstatt eines linearen Systems $(n + m + 1)$-ter Ordnung für a_i und b_i nur ein System m-ter Ordnung, um sämtliche Koeffizienten von $1/(x - x_i)$ in der „unvollständigen baryzentrischen Form" von Zähler und Nenner zu erhalten. Die Ausdehnung auf osculatorische Interpolation r-ter Ordnung durch $(a_0 + a_1x + \dots + a_{nr-1}x^{nr-1})/(1 + b_1x + \dots + b_{mr}x^{mr})$ erfordert die Lösung eines Systems rm-ter Ordnung anstelle des gewöhnlichen linearen Systems $(rm + rn)$-ter Ordnung. In ähnlicher Behandlung nicht-osculatorischer rationaler trigonometrischer Interpolation über eine ungerade Punktanzahl durch den Quotienten trigonometrischer Summen lösen wir ein lineares System $2m$-ter anstelle $(2n + 2m + 1)$-ter Ordnung für die Koeffizienten von $1/\sin \frac{1}{2}(x - x_i)$.

In rational interpolation of $f(x)$ by $(a_0 + a_1x + \dots + a_nx^n)/(1 + b_1x + \dots + b_mx^m)$ at x_i, $i = 1(1)\, n + m + 1$, instead of solving an $(n + m + 1)$-th order linear system for a_i and b_i, we solve only an m-th order linear system to obtain all $n + m + 1$ coefficients of $1/(x - x_i)$ in the "incomplete barycentric form" of the numerator and denominator. The extension to r-th order osculatory interpolation by $(a_0 + a_1x + \dots + a_{nr-1}x^{nr-1})/(1 + b_1x + \dots + b_{mr}x^{mr})$ requires the solution of an rm-th, instead of the usual $(rm + rn)$-th order linear system. In similar treatment of non-osculatory odd-point rational trigonometric interpolation by a quotient of trigonometric sums, we solve a $2m$-th, instead of $(2n + 2m + 1)$-th order linear system for the coefficients of $1/\sin \frac{1}{2}(x - x_i)$.

В рациональной интерполяции функции $f(x)$ по $(a_0 + a_1x + \dots + a_nx^n)/(1 + b_1x + \dots + b_mx^m)$ в точке x_i, $i = 1(1)\, n + m + 1$ вместо того, чтобы решать линейную систему $(n + m + 1)$-ого порядка для a_i и b_i, только решается линейная система m-ого порядка, чтобы получать все $n + m + 1$ коэффициенты $1/(x - x_i)$ в „неполной барицентрической форме" числителя и знаменателя. Распространение оскулаторной интерполации r-ого порядка требует решение линейной системы rm-ого порядка вместо обычной системы $(rm + rn)$-ого порядка. В подобной обработке неоскулаторной рациональной тригонометрической интерполяции с нечетным числом точек по частному тригонометрических сумм решается линейная система $2m$-ого порядка вместо системы $(2n + 2m + 1)$-ого порядка для коэффициентов от $1/\sin \frac{1}{2}(x - x_i)$.

I. Introduction

An explicit solution to the problem of $(n + m + 1)$-point rational interpolation of $f(x)$ by

$$\frac{N_n(x)}{D_m(x)} \equiv \frac{a_0 + a_1x + \dots + a_{n-1}x^{n-1} + a_nx^n}{b_0 + b_1x + \dots + b_{m-1}x^{m-1} + b_mx^m}, \tag{A}$$

i.e.,

$$N_n(x_i)/D_m(x_i) = f(x_i)\,, \qquad i = 1(1)\, n + m + 1\,, \tag{1}$$

was first given by CAUCHY [1]. His fractional form, which reduces to the LAGRANGE interpolation polynomial when $m = 0$, is far too unwieldy for practical use. Thus his numerator (denominator) has ${}_{n+m+1}C_{m+1}$ $({}_{n+m+1}C_m)$ terms, each a product of $m + 1$ (m) factors $f(x_i)$ and n (m) factors $(x - x_i)$ which is divided by $mn + n$ $(mn + m)$ factors $(x_i - x_j)$, for an offhand total, before possible regrouping, of ${}_{n+m+1}C_{m+1}(mn + 2n + m) + {}_{n+m+1}C_m(mn + 3m - 1) + 1$ multiplication-divisions.[1])

In preference to computation with CAUCHY's form, the prevailing practice has been to solve the $(n + m + 1)^{\text{th}}$ order linear system obtained from (1) for a_i and b_i (here normalized to $b_0 = 1$), and then to use (A) for computation.

We can determine a_i and b_i by solving first for the b_i's alone, which satisfy an mth order linear system derived from (1), since the m successive $(n + 1)^{\text{th}}$ order divided differences of $D_m(x)\, f(x)$ for the m sets of $n + 2$ nodes x_i, $x_{i+1}, \dots, x_{i+n+1}$, $i = 1(1)\, m$, must vanish. The elements of the augmented matrix of that system consist of m sets of $(n + 1)^{\text{th}}$ order divided differences of $x^kf(x)$, $k = 0(1)\, m$. Then the a_i in $N_n(x)$ may be best obtained from the nth degree LAGRANGE interpolation polynomial for $D_m(x)\, f(x)$ based upon any $n + 1$ of the $n + m + 1$ nodes x_i. Furthermore, this idea extends to osculatory rational interpolation by employing confluent divided differences.

In the present article we offer what may be a preferable alternative to the three preceding schemes. It may be less cumbersome and more direct, and it avoids all divided difference computations. At the same time it affords comparable reduction in the orders of the linear systems to be solved.

In Section II we show how an "incomplete barycentric form" of $N_n(x)$ and $D_m(x)$ gives an expression (B) that is equivalent to (A), and whose $n + m + 1$ coefficients of $1/(x - x_i)$, instead of a_i and b_i, may be found by solving just an mth order linear system, which is a substantial saving in computation when m is much less than n. In Sections III and IV we extend the idea in II to osculatory rational interpolation and non-osculatory rational trigonometric interpolation, by employing (C) and (E), having coefficients that may be found by solving linear systems whose orders are considerably less than the number of coefficients.

[1]) CAUCHY's formula is so rarely given, or even cited, that perhaps less than one person out of every thousand who can write down the LAGRANGE polynomial interpolation formula, can do the same for the former.

II. Rational interpolation

In (A), where $N_n(x_i)$ and $D_m(x_i)$ are presently unknown, express $N_n(x)$ and $D_m(x)$ as LAGRANGE interpolation polynomials for the respective sets of $n+1$ and $m+1$ nodes x_i, $i = 1(1)\,n+1$, and $x_0 = 0$, x_i, $i = n+2(1)\,n+m+1$, noting that $x_0 = 0$ is a necessary dummy node, since $D_m(0) = 1$ by the normalization. We have this equivalent form of (A):

$$N_n(x)/D_m(x) = P_{n+1}(x) \sum_{i=1}^{n+1} \frac{\alpha_i}{x - x_i} \Big/ Q_{m+1}(x) \sum_{i=0,n+2}^{n+m+1} \frac{\beta_i}{x - x_i}, \tag{B}$$

where

$$P_{n+1}(x) \equiv \Pi_{i=1}^{n+1} (x - x_i), \quad \text{and} \quad Q_{m+1}(x) \equiv x\, \Pi_{i=n+2}^{n+m+1} (x - x_i), \tag{2}$$

$$\alpha_i = N_n(x_i)/P'_{n+1}(x_i), \qquad i = 1(1)\,n+1, \tag{3}$$

$$\beta_i = D_m(x_i)/Q'_{m+1}(x_i), \qquad i = n+2(1)\,n+m+1, \tag{4}$$

and, since $D_m(0) = 1$,

$$\beta_0 = (-1)^m / \Pi_{j=n+2}^{n+m+1} x_j. \tag{4a}$$

The numerator and denominator of (B) are said to be in "*incomplete barycentric form*" because the term "*barycentric form*" is reserved for the expression of $P_{n+1}(x)$ and $Q_{m+1}(x)$ in (B) as

$$\left[\sum_{i=1}^{n+1} 1/P'_{n+1}(x_i)\,(x - x_i)\right]^{-1} \quad \text{and} \quad \left[\sum_{i=0,n+2}^{n+m+1} 1/Q'_{m+1}(x_i)\,(x - x_i)\right]^{-1}.$$

The advantage in (B) is that we have from it immediately an explicit expression for every α_i in terms of the β_i's, whereas in (A), to express every a_i in terms of the b_i's we must solve an $(n+1)^{\text{th}}$ order linear system for the $m+1$ different right members $(x_1^j, x_2^j, \ldots, x_{n+1}^j)^{\mathsf{T}}$, $j = 0(1)\,m$.

Multiplying (B) by $(x - x_i)\,Q_{m+1}(x)/P_{n+1}(x)$, $i = 1(1)\,n+1$, and then letting $x = x_i$, we obtain from (1)

$$\alpha_i = \left[\frac{(-1)^m}{x_i\, \Pi_{j=n+2}^{n+m+1} x_j} + \sum_{j=n+2}^{n+m+1} \frac{\beta_j}{(x_i - x_j)}\right] \frac{Q_{m+1}(x_i)\, f(x_i)}{P'_{n+1}(x_i)}, \qquad i = 1(1)\,n+1. \tag{5}$$

Putting α_i, given by (5), into (B) and multiplying now by $Q_{m+1}(x)/P_{n+1}(x)\,(x - x_j)$, $j = n+2(1)\,n+m+1$, and then letting $x = x_j$, which isolates β_j in the denominator of (B), we obtain from (1) the following linear system for β_j:

$$\sum_{k=n+2}^{n+m+1} \gamma_{jk}\beta_k = \varepsilon_j, \qquad j = n+2(1)\,n+m+1, \tag{6}$$

in which

$$\gamma_{jk} = S_j \sum_{i=1}^{n+1} R_i/(x_j - x_i)\,(x_i - x_k) - \delta_j^k, \tag{7}$$

$$\varepsilon_j = -S_j \sum_{i=1}^{n+1} R_i T_i/(x_j - x_i), \tag{8}$$

and where

$$S_j = P_{n+1}(x_j)/Q'_{m+1}(x_j)\, f(x_j), \tag{9}$$

$$R_i = Q_{m+1}(x_i)\, f(x_i)/P'_{n+1}(x_i), \tag{10}$$

$$\delta_j^k = 1, \quad k = j, \quad = 0, \quad k \neq j, \quad \text{and} \quad T_i = (-1)^m / x_i\, \Pi_{j=n+2}^{n+m+1} x_j. \tag{11}$$

After solving (6) for β_k, $k = n+2(1)\,n+m+1$, we find α_i, $i = 1(1)\,n+1$, from (5).

When n is considerably less than m, it is advantageous to invert (A), normalize by $a_0 = 1$, and apply (1) to $1/f(x)$, so that after expressing β_j, $j = 1(1)\,m+1$, in terms of α_i, $i = m+2(1)\,m+n+1$, we have to solve only n equations for the α_i's.

III. Osculatory rational interpolation

We may extend the idea in II to r^{th} *order osculatory rational interpolation* provided that *no set of osculatory interpolation conditions at any node is split between numerator and denominator*, which must therefore be of respective degrees $rn - 1$ and rm (the latter to accommodate the dummy node $x_0 = 0$ which is essential for the normalization $D_{rm}(0) = 1$). To meet the $rn + rm$ osculatory interpolation conditions at the $n + m$ distinct nodes x_i, i.e.,

$$\left[\frac{N_{rn-1}(x)}{D_{rm}(x)}\right]_{x=x_i}^{(s)} \equiv \left[\frac{a_0 + a_1x + \ldots + a_{rn-1}x^{rn-1}}{1 + b_1x + \ldots + b_{rm}x^{rm}}\right]_{x=x_i}^{(s)} = f^{(s)}(x_i), \quad i = 1(1)\,n+m, \quad s = 0(1)\,r-1, \tag{12}$$

requires the solution of an $(rn + rm)^{\text{th}}$ order linear system for a_i and b_i.[2]) But when $N_{rn-1}(x)$ and $D_{rm}(x)$ are expressed in complete barycentric form, obtaining their $rn + rm$ coefficients $\alpha_i^{(s)}$ and $\beta_i^{(s)}$ requires the solution of only an $(rm)^{\text{th}}$ order linear system. Here we define (cf. with (2))

$$P_n(x) \equiv \Pi_{i=1}^{n} (x - x_i), \qquad Q_m(x) \equiv \Pi_{i=n+1}^{n+m} (x - x_i), \tag{13}$$

and

$$P_{n,i}(x) \equiv \Pi_{j=1,j\neq i}^{n} (x - x_j), \qquad Q_{n,i}(x) \equiv \Pi_{j=n+1,j\neq i}^{n+m} (x - x_j). \tag{14}$$

[2]) Cf. remark following (C).

Corresponding to (B) in II we have here

$$\frac{N_{rn-1}(x)}{D_{rm}(x)} = \frac{[P_n(x)]^r \sum_{i=1}^n \sum_{s=1}^r \alpha_i^{(s)}/(x-x_i)^s}{x[Q_m(x)]^r\,[\beta_0/x + \sum_{i=n+1}^{n+m} \sum_{s=1}^r \beta_i^{(s)}/(x-x_i)^s]}, \tag{C}$$

in which $\beta_0 = (-1)^{mr}/[\Pi_{j=n+1}^{n+m} x_j]^r$.

We recall here an earlier lemma on osculatory rational interpolation on the equivalence of the set of conditions $[u(x)/v(x)]_{x=x_i}^{(s)} = f^{(s)}(x_i)$ to the set of conditions $u^{(s)}(x_i) = [f(x)\,v(x)]_{x=x_i}^{(s)}$ when $s = 0(1)\,r-1$ in both sets [2, pp. 487—488]. Applying the lemma to (12), we may multiply (C) by

$$(x-x_i)^r\, x[Q_m(x)/P_n(x)]^r\,[\beta_0/x + \textstyle\sum_{i=n+1}^{n+m} \sum_{s=1}^r \beta_i^{(s)}/(x-x_i)^s],$$

for $i = 1(1)\,n$, differentiate $r-s$ times, $s = 1(1)\,r$, and then set $x = x_i$, to obtain

$$\alpha_i^{(s)} = a_0(i,s)\,\beta_0 + \sum_{j=n+1}^{n+m} \sum_{t=1}^r a_{j,t}(i,s)\,\beta_j^{(t)}, \qquad i = 1(1)\,n, \qquad s = 1(1)\,r, \tag{15}$$

where

$$a_0(i,s) = \frac{1}{(r-s)!}\left[f(x)\left\{\frac{Q_m(x)}{P_{n,i}(x)}\right\}^r\right]_{x=x_i}^{(r-s)}, \tag{16}$$

and

$$a_{j,t}(i,s) = \frac{1}{(r-s)!}\left[\frac{xf(x)}{(x-x_j)^t}\left\{\frac{Q_m(x)}{P_{n,i}(x)}\right\}^r\right]_{x=x_i}^{(r-s)}. \tag{17}$$

By a proof similar to that for the lemma in [2], it is apparent that the sets of conditions $u^{(s)}(x_i) = f^{(s)}(x_i)$ and $[1/u(x)]_{x=x_i}^{(s)} = [1/f(x)]_{x=x_i}^{(s)}$, $s = 0(1)\,r-1$, are equivalent. Applying this to the remaining nodes in (12) by taking the reciprocals of (C) and $f(x)$, multiplying by

$$[(x-x_j)^r/x]\,[P_n(x)/Q_m(x)]^r \textstyle\sum_{i=1}^n \sum_{s=1}^r \alpha_i^{(s)}/(x-x_i)^s$$

for $j = n+1(1)\,n+m$, replacing $\alpha_i^{(s)}$ by the right member of (15), differentiating $r-s$ times, $s = 1(1)\,r$, and letting $x = x_j$, we obtain

$$\beta_j^{(s)} = \frac{1}{(r-s)!}\left[\frac{1}{xf(x)}\left\{\frac{P_n(x)}{Q_{m,j}(x)}\right\}^r \sum_{i=1}^n \sum_{t=1}^r \frac{a_0(i,t)\,\beta_0 + \sum_{k=n+1}^{n+m} \sum_{l=1}^r a_{k,l}(i,t)\,\beta_k^{(l)}}{(x-x_i)^t}\right]_{x=x_j}^{(r-s)},$$

$$j = n+1(1)\,n+m, \qquad s = 1(1)\,r. \tag{18}$$

We solve this $(rm)^{\text{th}}$ order linear system for $\beta_j^{(s)}$ (cf. with (6)):

$$\sum_{k=n+1}^{n+m} \sum_{l=1}^r \gamma_{j,k}^{(s,l)} \beta_k^{(l)} = \varepsilon_j^{(s)}, \qquad j = n+1(1)\,n+m, \qquad s = 1(1)\,r, \tag{19}$$

where

$$\gamma_{j,k}^{(s,l)} = \sum_{i=1}^n \sum_{t=1}^r S_{j,s}^{(i,t)} a_{k,l}(i,t) - \delta_s^l \delta_j^k, \tag{20}$$

and

$$\varepsilon_j^{(s)} = -\beta_0 \sum_{i=1}^r \sum_{t=1}^r S_{j,s}^{(i,t)} a_0(i,t), \tag{21}$$

in which

$$S_{j,s}^{(i,t)} = \frac{1}{(r-s)!}\left[\frac{1}{xf(x)\,(x-x_i)^t}\left\{\frac{P_n(x)}{Q_{m,j}(x)}\right\}^r\right]_{x=x_j}^{(r-s)}. \tag{22}$$

After solving (19) for $ß_j^{(s)}$, we obtain $\alpha_i^{(s)}$ from (15).[3]

For n considerably less than m, the remark on inversion at the end of II is more pertinent here, due to the greater margin between rn and rm.

When $m = 0$, where no equations have to be solved, $D_{rm}(x) = 1$, and $\beta_0 = 1$, $\beta_j^{(l)} = 0$ and $Q_m(x) = 1$ in (15) and (16), the $N_{rn-1}(x)$ in (C) is the incomplete barycentric from of HERMITE's $(rn-1)^{\text{th}}$ degree osculatory interpolation polynomial which, after applying LEIBNITZ's theorem to (16), may be rearranged to give this more explicit expression:

$$N_{rn-1}(x) = [P_n(x)]^r \sum_{i=1}^n \sum_{s=0}^{r-1} \left\{\sum_{j=s}^{r-1} \frac{[\{1/P_{n,i}(x)\}^r]_{x=x_i}^{(j-s)}}{s!\,(j-s)!\,(x-x_i)^{r-j}}\right\} f^{(s)}(x_i). \tag{23}$$

IV. Rational trigonometric interpolation

The idea in II applies, with slight modification, to *non-osculatory rational trigonometric interpolation* using an incomplete barycentric form of GAUSS's odd-point trigonometric interpolation formula [3]. For $(2n+2m+1)$-point interpolation of $f(x)$ by

$$\frac{T_n(x)}{T_m(x)} \equiv \frac{a_0 + \sum_{j=1}^n (a_j \cos jx + b_j \sin jx)}{c_0 + \sum_{j=1}^m (c_j \cos jx + d_j \sin jx)}, \tag{D}$$

i.e.,

$$T_n(x_i)/T_m(x_i) = f(x_i), \qquad i = 1(1)\,2n+2m+1, \tag{24}$$

[3]) Although the order of the linear system has been reduced, unlike the situation in II, here the total computational labor (considering in particular (16), (17) and (22), even though they are explicit) may very well come to considerably more than that in an $(rn+rm)^{\text{th}}$ order linear system from (12). Also, all reduced linear systems in this present article should be tested numerically, especially with regard to their condition.

we ordinarily solve (after suitable normalization of c_0) a $(2n + 2m + 1)^{\text{th}}$ order linear system for $a_0, a_r, b_r, r = 1(1)\,n$, and $c_r, d_r, r = 1(1)\,m$. To require only a $(2m)^{\text{th}}$ order system, write the numerator and denominator of (D) in this incomplete barycentric form, after first providing the dummy node $x_0 = 0$:

$$\frac{T_n(x)}{T_m(x)} = \frac{R_{2n+1}(x)\,\sum_{i=1}^{2n+1} \alpha_i/\sin \frac{1}{2}(x - x_i)}{S_{2m+1}(x)\,\sum_{i=0,\,2n+2}^{2n+2m+1} \beta_i/\sin \frac{1}{2}(x - x_i)}, \tag{E}$$

where

$$R_{2n+1}(x) \equiv \Pi_{j=1}^{2n+1} \sin \tfrac{1}{2}(x - x_j) \quad \text{and} \quad S_{2m+1}(x) \equiv \sin \tfrac{1}{2} x\, \Pi_{j=2n+2}^{2n+2m+1} \sin \tfrac{1}{2}(x - x_j)\,. \tag{25}$$

In place of the normalization $c_0 = 1$ for (D), it is more convenient to normalize (E) by $\beta_0 = 1$, that is $T_m(0) = \Pi_{i=2n+2}^{2n+2m+1} \sin \frac{1}{2} x_i$.[4])

Multiplying (E) by $\sin \frac{1}{2}(x - x_i)\,S_{2m+1}(x)/R_{2n+1}(x)$, $i = 1(1)\,2n + 1$, and letting $x = x_i$, we obtain from (24),

$$\alpha_i = \left[\frac{1}{\sin \frac{1}{2} x_i} + \sum_{j=2n+2}^{2n+2m+1} \frac{\beta_j}{\sin \frac{1}{2}(x_i - x_j)}\right] \frac{S_{2m+1}(x_i)\,f(x_i)}{\Pi_{j=1,\,j\neq i}^{2n+1} \sin \frac{1}{2}(x_i - x_j)}\,. \tag{26}$$

Putting these α_i's into (E) and multiplying now by

$$S_{2m+1}(x)/R_{2n+1}(x)\,\sin \tfrac{1}{2}(x - x_j)\,, \qquad j = 2n + 2(1)\,2n + 2m + 1\,,$$

and letting $x = x_j$, we obtain from (24)

$$\sum_{k=2n+2}^{2n+2m+1} \gamma_{jk}\beta_k = \varepsilon_j\,, \qquad j = 2n + 2(1)\,2n + 2m + 1\,, \tag{27}$$

where

$$\gamma_{jk} = S_j \sum_{i=1}^{2n+1} \frac{R_i}{\sin \frac{1}{2}(x_j - x_i)\,\sin \frac{1}{2}(x_i - x_k)} - \delta_j^k\,, \tag{28}$$

$$\varepsilon_j = -S_j \sum_{i=1}^{2n+1} \frac{R_i}{\sin \frac{1}{2} x_i \sin \frac{1}{2}(x_j - x_i)}, \tag{29}$$

and in (28) and (29),

$$S_j = \frac{R_{2n+1}(x_j)}{f(x_j)\,\sin \frac{1}{2} x_j\, \Pi_{k=2n+2,\,k\neq j}^{2n+2m+1} \sin \frac{1}{2}(x_j - x_k)}, \tag{30}$$

and

$$R_i = \frac{S_{2m+1}(x_i)\,f(x_i)}{\Pi_{k=1,\,k\neq i}^{2n+1} \sin \frac{1}{2}(x_i - x_k)}\,. \tag{31}$$

After solving (27) for β_j, we obtain α_i from (26).

There appears to be difficulty in finding something similar to III to reduce the number of equations, for osculatory rational trigonometric interpolation, because we cannot express a trigonometric sum, having the same number of coefficients as interpolation conditions, in an incomplete barycentric form that involves successive powers of $1/\sin \frac{1}{2}(x - x_i)$ beyond the first.

References

1 CAUCHY, L., Sur la formule de LAGRANGE relative à l'Interpolation, Cours d'Analyse de l'Ecole Royale Polytechnique (Analyse Algébrique), Note V, Paris, 1821, repr. in Oeuvres, ser. 2, vol. III, 1897, pp. 429–433.
2 SALZER, H. E., Note on osculatory rational interpolation, Math. Comput. **16** (1962), 486–491.
3 SALZER, H. E., Coefficients for facilitating trigonometric interpolation, J. Math. Phys. **27**, (1949), 274–278.

Eingereicht: 15. 2. 1980, revidierte Fassung 13. 9. 1980

Anschrift: Professor Dr. HERBERT E. SALZER, 941 Washington Avenue, Brooklyn, N.Y. 11225, U.S.A.

[4]) The normalization $T_m(0) = 1$, that is, $\beta_0 = 1/\Pi_{i=2n+2}^{2n+2m+1} \sin \frac{1}{2} x_i$, would resemble (4a) which seems to be more natural for $D_m(x)$ than $T_m(x)$.

ZAMM 61, 165–173 (1981)

R. FINN

On the Laplace Formula and the Meniscus Height for a Capillary Surface

Die Höhe im Mittelpunkt und die „Meniskushöhe" einer symmetrischen Kapillarfläche werden von oben und von unten abgeschätzt. Die Resultate sind für enge Röhren asymptotisch genau und behalten auch für breitere Röhren eine Bedeutung.

Upper and lower bounds are given for the center and meniscus heights of a symmetric equilibrium capillary surface. The bounds are sensibly identical at small Bond numbers B, and they provide significant information also for intermediate values of B.

Получены верхные и нижние оценки для высоты симметричной капиллярной поверхности в центре и на крае. Оценки асимптотически точны для узких трубок, и имеют смысл также и для широких трубок.

In 1805, PIERRE SIMON LAPLACE [1] presented the differential equation

$$\operatorname{div} \frac{1}{\sqrt{1 + |\nabla u|^2}} \nabla u = \varkappa u \tag{1}$$

for the height $u(x, y)$ of the free surface $\mathcal{S}$ in a cylindrical capillary tube of arbitrary section in an infinite reservoir (fig. 1). Here $\varkappa = \varrho g/\sigma$ is the "capillarity constant", ϱ the density difference across the surface, g the gravitational acceleration, σ the surface tension. In the configuration illustrated, $\varkappa > 0$. In the same work, LAPLACE integrated (1) approximately in the case of circular section of small radius a, to obtain the celebrated formula

$$u_0 \sim 2 \frac{\cos \gamma}{\varkappa a} - \frac{a}{\cos \gamma} \left(1 - \frac{2}{3} \frac{1 - \sin^3 \gamma}{\cos^2 \gamma}\right) \tag{2}$$

for the height u_0 on the axis of symmetry, on the condition that $\mathcal{S}$ meet the cylinder walls Z in a prescribed constant angle γ.

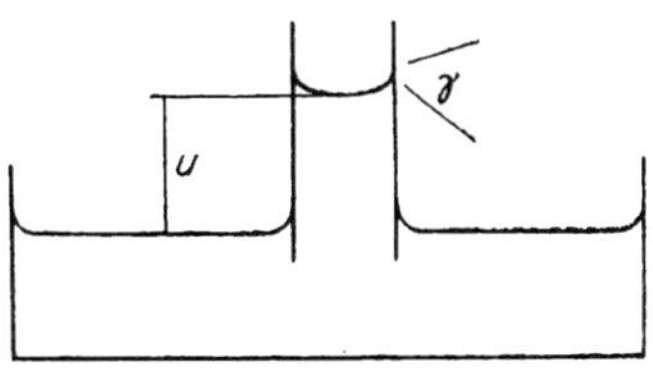

Fig. 1

LAPLACE discovered the relation (2), he did not prove it, nor did he offer error bounds for its accuracy. The formula is, however, asymptotically correct, as was proved by DAVID SIEGEL [2] in 1978. SIEGEL's proof can be made to yield error bounds, although these bounds become (on one side) non-uniform as $\gamma \to 0$.

We present in this note another proof of LAPLACE's formula, which is simpler than the one given by SIEGEL, and which leads to improved bounds. We show in particular that *on the axis of symmetry the exact solution of* (1) *lies in every case strictly above the estimate* (2) *of* LAPLACE.

Our procedure leads also to bounds above and below for the *meniscus height* q being the difference between maximum and minimum heights. For the case of narrow tube considered here, expressions for approximating u_0 and q are given, i.a., in LAPLACE [1], POISSON [3], RAYLEIGH [4] and CONCUS [5]. Those references do not provide rigorous error estimates. In Tables 1 and 2, and in Figures 3 and 4, we compare the results of the authors just cited with the permissible ranges for u_0 and q, as determined in the present study.

The underlying basis for our procedure is an explicit formula (essentially (5)) going back to LAPLACE, which expresses the volume of fluid lifted by the capillary surface above the reference level, in terms of the contact angle and the radius of the tube. For narrow tubes, our estimates follow on a judicious choice of two spherical "comparison" surfaces, which yield, respectively, smaller and larger lifted volumes.

For larger tubes, we apply general estimates for u_0 and for q, derived in our earlier work [6] for symmetric tubes of arbitrary diameter. The results of the present paper improve those estimates in the special case of narrow tubes; from that point of view, they can be considered as a completion of [6] for a situation that is not fully developed in that reference.

We remark that the present results for symmetric surfaces can be used in conjunction with general comparison procedures (some of which are idiosyncratic for the nonlinearity in the equation) to obtain bounds above and below for capillary surfaces in tubes of general cross section. For studies in that direction, we refer the reader to our earlier papers [6, 7], to the work of SIEGEL [2], and to the references cited in those papers.

1. The solutions of (1) with rotational symmetry are described by the ordinary differential equation

$$(r \sin \psi)_r = \varkappa r u \tag{3}$$

where $\psi = \tan^{-1} u'(r)$ is the angle made with the horizontal by a meridian curve. We observe that (1) (or (3)) expresses geometrically that the mean curvature H of the surface $u(x, y)$ is proportional to u, and we write (3) in the form

$$\frac{\sin\psi}{r} + (\sin\psi)_r \equiv k_l + k_m \equiv 2H = \varkappa u\,, \tag{4}$$

thus splitting H into latitudinal and meridional sectional curvatures k_l and k_m.

We have from (3) the explicit formula

$$r \sin\psi = \varkappa \int_0^r \varrho u(\varrho)\, d\varrho \tag{5}$$

for (essentially) the volume of fluid lifted above reference level by the capillary surface. From (5) follows that if $u_0 > 0$, then $u(r)$ is increasing, and hence, putting its initial and final values into (5)

$$\varkappa \frac{u_0}{2} < \frac{\sin\psi}{r} < \varkappa \frac{u}{2} \tag{6a, b}$$

on any interval into which the solution can be continued.

From (4) we now find, first,

$$(\sin\psi)_r > \frac{\varkappa}{2} u(r) > \frac{\varkappa}{2} u_0 > 0$$

so that

$$k_m = (\sin\psi)_r > 0 \tag{7}$$

and consequently $u_{rr} > 0$; thus,

$$\frac{d}{dr} k_m = (\sin\psi)_{rr} = \varkappa u_r - \frac{1}{r}(\sin\psi)_r + \frac{\sin\psi}{r^2} = \varkappa u_r - \frac{1}{r}\left(\varkappa u - \frac{\sin\psi}{\varkappa r}\right) + \frac{\sin\psi}{r^2}$$

$$= \varkappa u_r - \frac{\varkappa}{r}\left(u - 2\frac{\sin\psi}{\varkappa r}\right) > \varkappa u_r - \frac{\varkappa}{r}(u - u_0)$$

by (6a), so that

$$\frac{d}{dr} k_m > \frac{\varkappa}{r} \int_0^r \varrho u_{\varrho\varrho}(\varrho)\, d\varrho > 0\,. \tag{8}$$

Similarly, we find

$$\frac{d}{dr} k_l = \frac{d}{dr}\frac{\sin\psi}{r} = \frac{(\sin\psi)_r}{r} - \frac{\sin\psi}{r^2} = \frac{1}{r}\left(\varkappa u - \frac{\sin\psi}{r}\right) > 0 \tag{9}$$

by (6b). Thus, *both sectional curvatures increase monotonically with r.*

Returning to (4), we note $(\sin\psi)_r = -(\cos\psi)_u$; thus, using (6),

$$\varkappa \frac{u}{2} < -(\cos\psi)_u < \varkappa\left(u - \frac{u_0}{2}\right) \tag{10}$$

from which

$$\frac{u_0}{2} + \left(\frac{2}{\varkappa}(1 - \cos\psi) + \frac{u_0^2}{4}\right)^{1/2} < u(r) < \left(\frac{4}{\varkappa}(1 - \cos\psi) + u_0^2\right)^{1/2}. \tag{11a, b}$$

2. *Let $u_1(r)$, $u_2(r)$ define symmetric surfaces over a disk $0 \leq r < R$; suppose $u_1(0) = u_2(0)$, and suppose the mean curvatures $H_1(r)$, $H_2(r)$ satisfy $H_1 \geqq H_2$. Then $u_1(r) \geqq u_2(r)$, $u_1'(r) \geqq u_2'(r)$ in $0 \leqq r < R$. Equality holds at $r_0 > 0$ if and only if $H_1(r) = H_2(r)$, all $r < r_0$.*

Proof: $(r \sin\psi_1)_r = 2H_1(r)$, $(r \sin\psi_2)_r = 2H_2(r)$, hence

$$r(\sin\psi_1 - \sin\psi_2) = 2\int_0^r \varrho(H_1 - H_2)\, d\varrho;$$

thus $\psi_1 \geqq \psi_2$, which implies $u_1'(r) \geqq u_2'(r)$. Since $u_1(0) = u_2(0)$, the result follows.

The above result, in conjunction with the material of 1., shows that *the set of all symmetric capillary surfaces in tubes of circular section, with $0 \leqq \gamma \leqq \pi/2$, corresponds biuniquely to the set of all choices of initial value $u_0 \geqq 0$.*

3. Let $u(r; u_0)$ satisfy (3), with $u(0; u_0) = u_0 > 0$. The mean curvature of the (capillary) surface defined by u is then $H(r) = \frac{1}{2}\varkappa u$; by (6, 11) there holds

$$\frac{\varkappa}{2} u_0 < H(r) < \frac{\varkappa}{2}\left(\frac{4}{\varkappa}(1 - \cos\psi) + u_0^2\right)^{1/2}$$

on the interval indicated. It follows then from **2.** that the lower hemisphere Σ_1: $v_1(r; u_0)$ of radius $R_1 = 2/\varkappa u_0$, centered on the axis $r = 0$ and such that $v_1(0; u_0) = u_0$, will satisfy $v_1(r; u_0) < u(r; u_0)$ in the interval $r > 0$ for which $u(r; u_0)$ is defined (see Fig. 2).

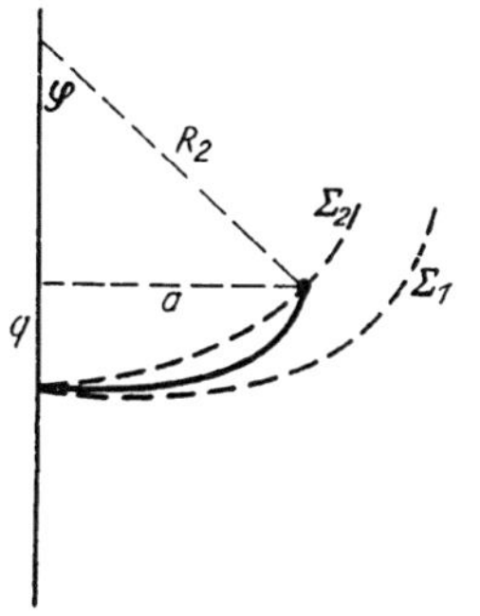

Fig. 2

Let $(a; u(a; u_0)) = p_a$ be a point on a solution section. The relation (11), together with the inequality

$$\left(\frac{4}{\varkappa}(1 - \cos\psi) + u_0^2\right)^{1/2} < u_0 + \frac{2}{\varkappa u_0}(1 - \cos\psi)$$

shows that p_a lies strictly interior to the circular segment determined by Σ_1 up to the height $u_0 + \frac{2}{\varkappa u_0}(1 - \cos\psi)$. Thus, *the lower hemisphere $v_2(r; u_0)$, centered on the axis $r = 0$, with $v_2(0; u_0) = u_0$, $v_2(a; u_0) = u(a; u_0)$, has radius $R_2 < R_1$.*

4. *We assert that $u < v_2$ on the entire interval* $0 < r < a$. Also, *if φ is the inclination angle of the curve $v_2(r; u_0)$, then $\varphi(a) < \psi(a)$.*

In fact, we observe first that $R_2 < R_1 = 2/\varkappa u_0$, which is the (planar) curvature k_m of $u(r; u_0)$ at $r = 0$. Thus, the assertion holds on some initial interval. Let $r_2 \leqq a$ be the first point of contact. By **2.** there must be a subinterval in which $k_m > 1/R_2$. But we have shown in **1.** that k_m is increasing; thus $k_m(r_2) > 1/R_2$, and hence $u > v_2$ for $r > r_2$. We conclude $r_2 = a$, as was to be shown. The remainder of the statement is now immediate.

We note for later reference the Consequence

$$R_2 = \frac{a}{\sin\varphi(a)} > \frac{a}{\sin\psi(a)} \tag{12}$$

of our result.

5. We may now adopt $v_2(r; u_0)$ as an upper comparison surface on $0 \leqq r \leqq a$. We have, by (3) and the above,

$$a \sin\psi = \varkappa \int_0^a r u(r; u_0)\, dr < \varkappa \int_0^a r v_2(r; u_0)\, dr \tag{13}$$

with

$$v_2(r; u_0) = u_0 + R_2 - (R_2^2 - r^2)^{1/2}. \tag{14}$$

Integration yields

$$u_0 > \frac{2\sin\psi}{\varkappa a} - R_2 + \frac{2}{3a^2}[R_2^3 - (R_2^2 - a^2)^{3/2}]. \tag{15}$$

One verifies that the right side of (15) is monotone increasing in R_2. By (12) we have

$$R_2 > \frac{a}{\sin\psi}.$$

We conclude

$$u_0 > L(a; \psi) \equiv \frac{2\sin\psi}{\varkappa a} - \frac{a}{\sin\psi} + \frac{2a}{3\sin^3\psi}(1 - \cos^3\psi). \tag{16}$$

The right side of (16) is exactly the Laplace expression (2). Thus, *the* Laplace *formula provides a strict lower bound for the center height u_0 in a circular capillary tube.*

6. We may obtain an upper bound by using the function $v_1(r)$ as (lower) comparison surface. We proceed as above, obtaining now

$$\frac{a}{\varkappa}\sin\psi > \frac{a^2 u_0}{2} + \frac{a^2}{2}R_1 - \frac{1}{3}[R_1^3 - (R_1^2 - a^2)^{3/2}] \tag{17}$$

with $R_1 = 2/\varkappa u_0$.

This relation holds for any solution surface, hence the actual u_0 is less than the least upper bound of values u_0 that satisfy (17). The right side of (17) is monotone increasing in u_0, hence there is a unique $u_0 = u_0^+$ for which the inequality is replaced by equality; this value serves as an upper bound for u_0. Thus, *there holds always* $u_0 < u_0^+$, *where* u_0^+ *is the unique non-zero solution of the equation*

$$u_0^4 - \frac{2}{\varkappa a}u_0^3\sin\psi + \frac{2}{\varkappa}u_0^2 - \frac{16}{3\varkappa^3 a^2}\left[1 - \left(1 - \frac{\varkappa^2 a^2}{4}u_0^2\right)^{3/2}\right] = 0\,. \tag{18}$$

Asymptotically as $a \to 0$, one finds

$$u_0^+ \sim L(a;\psi) + \frac{\varkappa}{6}a^3\frac{\sin\psi}{(1+\cos\psi)^4}(1 + 2\cos\psi)\,. \tag{19}$$

7. We apply now the preceding methods and results to obtain upper estimates for the "meniscus height"

$$q(a;\psi) = u(a) - u_0\,.$$

We have immediately

$$q = R_2(1 - \cos\varphi) = \frac{a}{\sin\varphi}(1 - \cos\varphi) \tag{20}$$

(see Fig. 2), and thus, since $\varphi < \psi$,

$$q < \frac{a}{\sin\psi}(1 - \cos\psi) = \frac{a\sin\psi}{1 + \cos\psi} \tag{21}$$

which provides an initial (primitive) estimate.

A result that is in general (although not in every case) more precise is obtained by using (11b), from which follows

$$q < \frac{4(1 - \cos\psi)}{\varkappa(u + u_0)}\,. \tag{22}$$

Here we estimate u_0 from (16) and u either from (11a) or (more precisely) by using the result

$$u > \left(\frac{p+1}{p}\right)^{1/2}\left\{\frac{2}{\varkappa}(1 - \cos\psi) + \frac{p+1}{4}u_0^2\right\}^{1/2} \tag{23}$$

with

$$p = \left\{1 + \frac{2\varkappa a^2}{1 + \cos\psi}\right\}^{1/2} \tag{24}$$

taken from our earlier paper [6].

A still preferable procedure is to start with the inequality (22) of [6][1])

$$u < \frac{\sin\psi}{\varkappa a} + \left\{\frac{2}{\varkappa}(1 - \cos\psi) - \frac{\sin^2\psi}{\varkappa^2 a^2} + \frac{u_0^2}{2}\right\}^{1/2} \tag{25}$$

from which

$$q < \frac{\sin\psi}{\varkappa a} + \left\{\frac{2}{\varkappa}(1 - \cos\psi) - \frac{\sin^2\psi}{\varkappa^2 a^2} + \frac{u_0^2}{2}\right\}^{1/2} - u_0\,. \tag{26}$$

Denoting the right side of (26), for fixed ψ, a, by $f(u_0)$, we find

$$f'(u_0) = \lambda\left(u_0 - 2\left\{\frac{2}{\varkappa}(1 - \cos\psi) - \frac{\sin^2\psi}{\varkappa^2 a^2} + \frac{u_0^2}{2}\right\}^{1/2}\right) \tag{27}$$

with $\lambda > 0$. We assert first that *on any solution curve there holds* $f'(u_0) < 0$. We have, in fact, by (6b)

$$\left(u - \frac{\sin\psi}{\varkappa a}\right)^2 > \left(\frac{\sin\psi}{\varkappa a}\right)^2 \tag{28}$$

---

[1]) See also SIEGEL [2], where the inequality first appears.

and thus, from (25)

$$u_0^2 > -\frac{4}{\varkappa}(1-\cos\psi) + \frac{4\sin^2\psi}{\varkappa^2 a^2}. \tag{29}$$

Inserting (29) into (27) and using (6a) yields

$$f'(u_0) < \frac{2\sin\psi}{\varkappa a} - 2\left(\frac{\sin^2\psi}{\varkappa^2 a^2}\right)^{1/2} = 0 \tag{30}$$

which was to be shown.

We shall need to consider value sets (u_0, a, ψ) that need not lie on solution curves, and we thus seek the range of u_0 for which (30) continues to hold. Equality holds if and only if

$$u_0^2 = -\frac{8}{\varkappa}(1-\cos\psi) + \frac{4\sin^2\psi}{\varkappa^2 a^2}; \tag{31}$$

since (29) holds on the solution curve, we conclude $f'(u_0) < 0$ for all u_0 that satisfy

$$u_0^2 > -\frac{8}{\varkappa}(1-\cos\psi) + \frac{4\sin^2\psi}{\varkappa^2 a^2}. \tag{32}$$

But (32) holds as consequence of (16), *hence we may use* (16) in (26) to estimate q *from above.*

8. A lower bound for q analogous to the upper estimate (25) can be obtained by using the lower comparison sphere Σ_1; the result obtained that way is however not very satisfactory. A preferable procedure is to start with (11a), which yields

$$q > -\frac{u_0}{2} + \left\{\frac{u_0^2}{4} + \frac{2}{\varkappa}(1-\cos\psi)\right\}^{1/2} \tag{32}$$

and to observe that the right side of (33) is decreasing in u_0. Thus, an estimate is obtained by substituting (6a) or (more accurately) the solution of (18) for u_0.

A general comparison with the upper bound is obtained by noting the inequality

$$\sqrt{1+x} > 1 + \tfrac{1}{2}x - \tfrac{1}{8}x^2$$

which holds in $0 \leqq x \leqq 8$. Thus,

$$q > \frac{2}{\varkappa u_0}(1-\cos\psi) - \frac{4}{\varkappa^2 u_0^3}(1-\cos\psi)^2 \tag{33}$$

whenever[2])

$$\varkappa u_0^2 > (1-\cos\psi). \tag{34}$$

A still more precise (if less tractable) lower bound follows from (23): thus

$$q > \left(\frac{p+1}{p}\right)^{1/2}\left\{\frac{2}{\varkappa}(1-\cos\psi) + \frac{p+1}{4}u_0^2\right\}^{1/2} - u_0. \tag{35}$$

We show that in general *this inequality persists if* u_0 *is replaced by the solution of* (18). To do so, we denote the right side of (35) (for fixed ψ, p) by $f(u_0)$, and observe that $f'(0) < 0$. At any zero of $f'(u_0)$, there holds

$$u_0^2 = \frac{16p}{p-1}\frac{\sin^2\psi}{\varkappa^2 a^2}. \tag{36}$$

Thus, $f(u_0)$ is decreasing until the value of u_0 indicated by (36). This value however exeeds the solution of (18), hence (35) continues to hold if u_0 is replaced by u_0^+.

9. For convenience of reference we restate here some of our estimates in nondimensional form in terms of the variables

$$U = \frac{u}{a}, \qquad Q = \frac{q}{a}, \qquad B = \varkappa a^2 = \frac{\varrho g}{\sigma} a^2$$

and boundary angle

$$\gamma = \frac{\pi}{2} - \psi.$$

---

[2]) According to (11b), (6a), it will suffice to have $\varkappa a^2 < (4/5)(1+\cos\psi)$.

Writing

$$\mathscr{L}(B;\gamma) \equiv 2B^{-1}\cos\gamma - \frac{1}{\cos\gamma}\left(1-\frac{2}{3}\frac{1-\sin^3\gamma}{\cos^2\gamma}\right) \equiv 2B^{-1}\cos\gamma - \frac{(1+2\sin\gamma)\cos\gamma}{3(1+\sin\gamma)^2} \tag{37}$$

and

$$\mathscr{N}(B;P;\gamma) \equiv 2B^{-1}\cos\gamma - P + \tfrac{2}{3}[P^3-(P^2-1)^{3/2}] \tag{38}$$

we find

$$U_0 > \mathscr{N}_0 = \mathscr{N}\left(B;\frac{1}{\cos\gamma};\gamma\right) \equiv \mathscr{L}(B;\gamma) \tag{39}$$

and

$$U_0 < U_0^+ \tag{40}$$

where U_0^+ is the unique solution of

$$U_0 - \mathscr{N}\left(B;\frac{2}{BU_0};\gamma\right) = 0\,. \tag{41}$$

It is convenient to write (41) in terms of the auxiliary variable $P_0 = 2/BU_0$. We find

$$P_0^{-1} - \cos\gamma + \tfrac{1}{2}BP_0 - \tfrac{1}{3}B[P_0^3-(P_0^2-1)^{3/2}] = 0\,. \tag{42}$$

This relation is monotonic in P_0, hence easily solved.

We have always

$$Q < \frac{\cos\gamma}{1+\sin\gamma}\,; \tag{43}$$

more precisely,

$$Q < \frac{\cos\gamma}{B} + \left\{\frac{2}{B}(1-\sin\gamma) - \frac{\cos^2\gamma}{B^2} + \frac{1}{2}\mathscr{N}_0^2\right\}^{1/2} - \mathscr{N}_0$$

whenever $\mathscr{N}_0$ satisfies

$$\mathscr{N}_0^2 > -\frac{8}{B}(1-\sin\gamma) + \frac{4\cos^2\gamma}{B^2}\,. \tag{44}$$

We have also, from (35, 24, 36)

$$Q > \left(\frac{p+1}{p}\right)^{1/2}\left\{\frac{2}{B}(1-\sin\gamma) + \frac{p+1}{4}U_0^{+2}\right\}^{1/2} - U_0^+ \tag{45}$$

with

$$p = \left\{1+\frac{2B}{1+\sin\gamma}\right\}^{1/2} \tag{46}$$

whenever U_0^+ satisfies

$$U_0^+ < 4\left(\frac{p}{p-1}\right)^{1/2}\frac{\cos\gamma}{B}\,. \tag{47}$$

10. An alternative lower bound for U_0 can be obtained by using the estimate (44) of our earlier paper [6]. In the present nondimensional form, we obtain

$$U_0 > 4\frac{\cos\gamma}{B}\,\frac{1+\sqrt{1+B}}{1+\sin\gamma+\sqrt{2(1+\sin\gamma)}}\exp\left\{2\cos\left(\frac{\pi}{4}-\frac{\gamma}{2}\right)-\sqrt{1+B}-1\right\}. \tag{48}$$

The original choice (39) is preferable for small values of B. (In the case $\gamma = 0$, the "crossover point" is $B = 2.889$.)

11. The formal estimates of LAPLACE [1] have been rederived and developed further, from varying points of view, by POISSON [3], by RAYLEIGH [4], and by CONCUS [5], among others. When transcribed to our (nondimensional) notation, the formulas given by these authors become as follows:

1) LAPLACE:

$$U_0 \sim \frac{2\cos\gamma}{B} - \frac{1+2\sin\gamma}{3(1+\sin\gamma)^2}\cos\gamma\,,$$

$$Q \sim \frac{2}{BU_0}\left(1-\frac{2}{BU_0^2}\right)(1-\sin\gamma) + \frac{2}{3}\frac{1}{BU_0^2}(1+\cos 2\gamma) + \frac{8}{3}\frac{1}{BU_0^2}\log\cos\left(\frac{\pi}{4}-\frac{\gamma}{2}\right).$$

2) Poisson (for $\cos\gamma \gg 0$):

$$U_0 \sim \frac{2\cos\gamma}{B} - \lambda + \frac{2}{3}\lambda^3 - \frac{2}{3}(\lambda^2-1)^{3/2} - \frac{B}{3}\lambda^3\left(\lambda^2 - \lambda\sqrt{\lambda^2-1} - \frac{1}{2} + \log\frac{\lambda+\sqrt{\lambda^2-1}}{2\lambda}\right),$$

$$Q \sim \lambda - \sqrt{\lambda^2-1} + \frac{B}{3}\lambda^3 \log\frac{\lambda+\sqrt{\lambda^2-1}}{2\lambda} \quad \text{with} \quad \lambda = \frac{1}{\cos\gamma} - \frac{B}{3\cos^5\gamma}(1-\sin\gamma)\sin^2\gamma\,.$$

Table 1. Upper and lower bounds for $BU_0 = Bu_0/a$

$\gamma\downarrow$ $B\rightarrow$	.001	.01	.1	1	10
0	1.9996668 1.9996667	1.996682 1.996667	1.96800 1.96667	1.7413 1.6667	.87231 .3925*
$\pi/6$	1.7317943 1.7317942	1.729491 1.729485	1.70695 1.70639	1.5183 1.4755	.75909 .3490*
$\pi/3$	.99986925 .99986923	.9986942 .9986923	.987111 .986923	.8854 .8692	.44240 .2094*

* Computed from (48).

Table 2. Upper and lower bounds for Q

$\gamma\downarrow$ $B\rightarrow$	.001	.01	.1	1	10
0	.99981 .99967	.99807 .99669	.9818 .9686	.8789 .7927	.4975 .4193
$\pi/6$	.57731 .57724	.57695 .57628	.5735 .5670	.5475 .4979	.3568 .2917
$\pi/3$	.26794 .26791	.26785 .26758	.2669 .2644	.2598 .2390	.1856 .1492

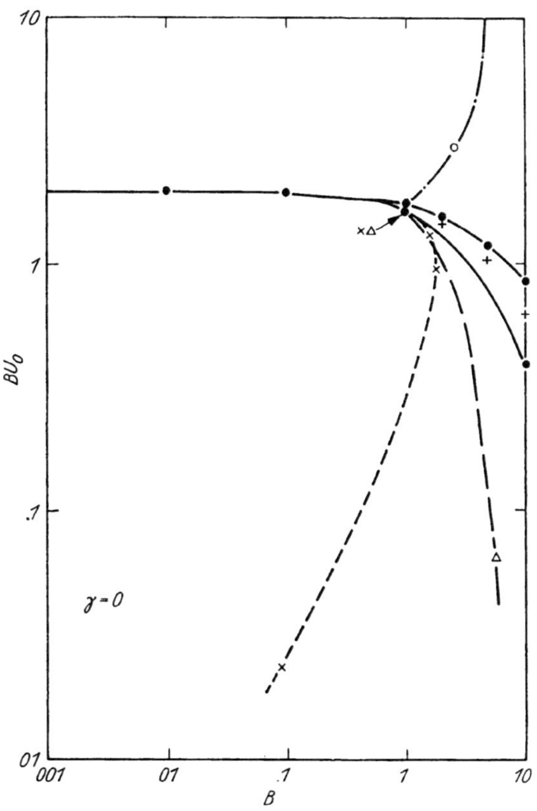

3a

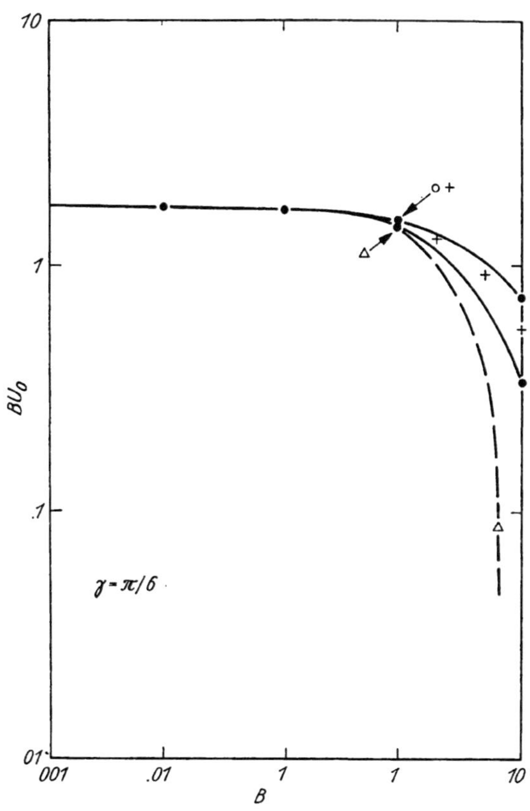

3b

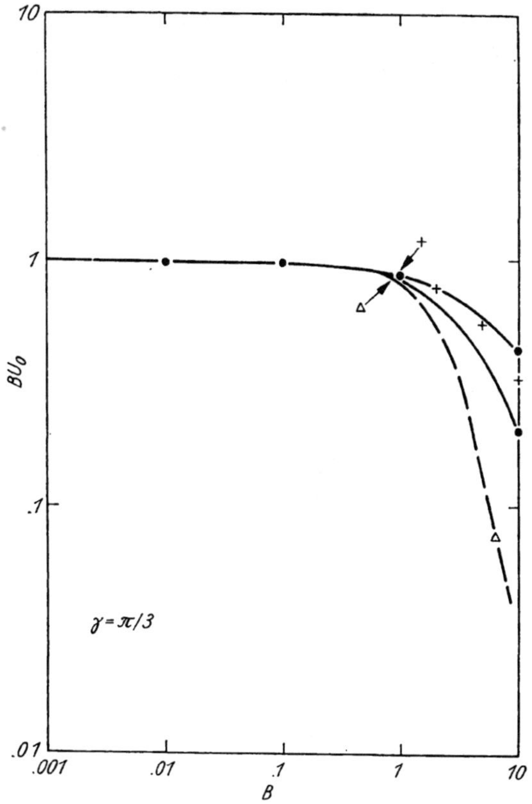

3c

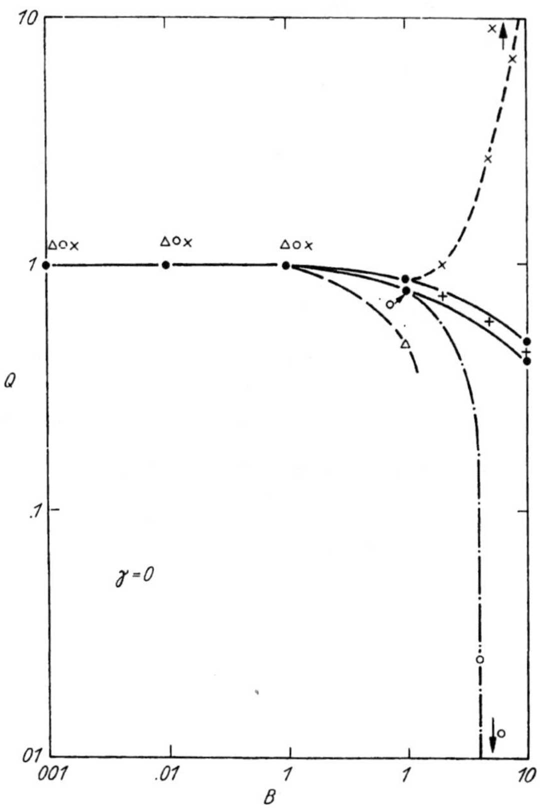

4a

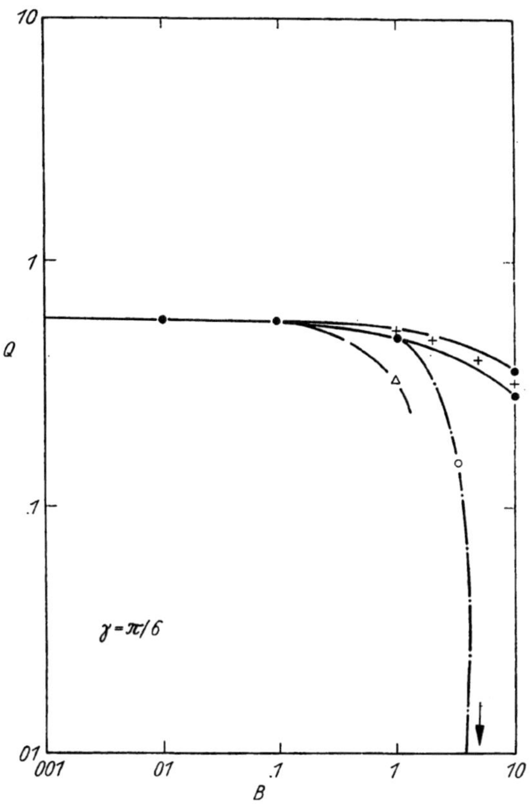

4b

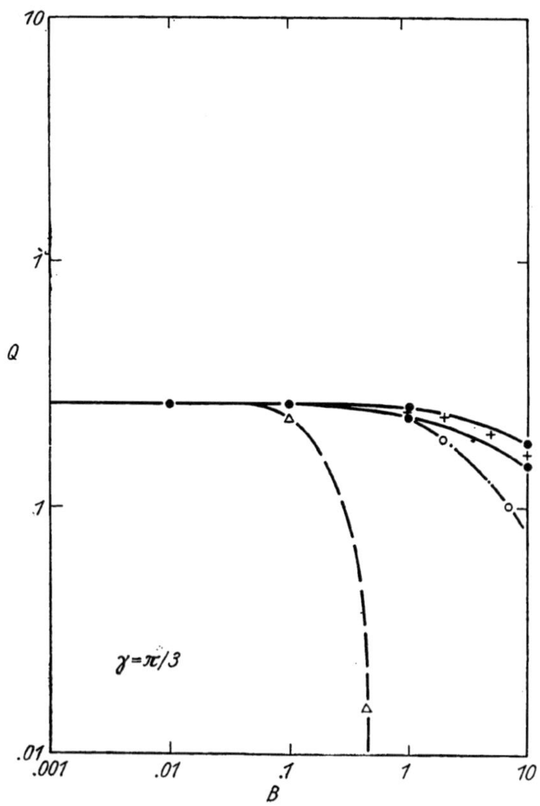

4c

Fig. 3. Normalized center heights for varying BOND number
[• Upper and lower bounds (solid lines); △ LAPLACE, CONCUS; ○ POISSON; × RAYLEIGH; + Computer calculations (CONCUS)]

Fig. 4. Meniscus heights for varying BOND number
[• Upper and lower bounds (solid lines); △ LAPLACE; ○ POISSON, CONCUS; × RAYLEIGH; + Computer calculations (CONCUS)]

3) RAYLEIGH (for $\cos\gamma = 1$):

$$\frac{2}{B} \sim U_0 + \frac{1}{3} - \frac{2}{3U_0}\left(\log 2 - \frac{1}{2}\right) + \frac{1}{9U_0^2}(32\log 2 - 21)\,,$$

$$Q \sim 1 - \frac{B}{3}\left(1 - \frac{B}{2}\right)\log 2\,.$$

4) CONCUS[3]):

$$U_0 \sim \frac{2\cos\gamma}{B} - \frac{1 + 2\sin\gamma}{3(1+\sin\gamma)^2}\cos\gamma\,,$$

$$Q \sim \frac{\cos\gamma}{1+\sin\gamma} + \frac{B}{\cos^3\gamma}\left[-\frac{1}{6} + \frac{\sin\gamma}{2} - \frac{\sin^2\gamma}{3} + \frac{1}{3}\log\left(\frac{1+\sin\gamma}{2}\right) + \lambda\,\frac{1-\sin\gamma}{2}\cos\gamma\right]$$

with

$$\lambda = \frac{1}{\cos\gamma} - \frac{2}{3}\,\frac{1-\sin^3\gamma}{\cos^3\gamma} = \frac{1}{3}\,\frac{1+2\sin\gamma}{(1+\sin\gamma)^2}\cos\gamma\,.$$

In Tables 1 and 2, and in Figures 3 and 4, we compare the values predicted by the above cited authors, with the ranges of U_0 and of Q as given in the present work, for varying choices of B and of γ. We note that values $B \gg 0.1$ are generally considered too large to permit satisfactory calculation with asymptotic formulae (cf. CONCUS [5, p. 492]).

12. One sees that the results of the present study, besides giving explicit upper and lower bounds, provide useful information at least to a value $B \approx 10$ (which, according to the calculations of CONCUS, lies in the upper range of the "intermediate" size configurations). It seems likely that the results can be made still more precise, and perhaps extended throughout the "intermediate" range, by using the present method in conjunction with an improved choice of comparison surfaces. Estimates (asymptotically exact) for large B appear in our earlier work [6] where they are developed for the case of a sessile drop on a horizontal surface. An example worked out in that paper (§ 8.3), for a value $B \approx 750$ in a more general situation, yielded an estimate for Q accurate to within 0.6%.

Note added in proof: The conditions (44), (47) are superfluous when U_0 is estimated by (39), (40). See the comments following (32), (36).

References

1 LAPLACE, P. S., Traité de mécanique céleste, Tome quatrième, Supplément au livre X, „Sur l'action capillaire", Gauthier-Villars, Paris 1806. See also the annotated translation by N. BOWDITCH, vol. IV, Chelsea Publishing Co., Bronx, N.Y., 1966.
2 SIEGEL, D., Height estimates for capillary surfaces, Pacific J. Math., to appear.
3 POISSON, S. D., Nouvelle théorie de l'action capillaire, Bachelier père et fils, Paris 1831.
4 RAYLEIGH, J. W. S., On the theory of the capillary tube, Proc. Roy. Soc. London, Ser. A 92 (1915), 184—195.
5 CONCUS, P., Static menisci in a vertical right circular cylinder, J. Fluid Mech. 34 (1968), 481—495.
6 FINN, R., The sessile liquid drop. I: symmetric case, Pacific J. Math., to appear.
7 FINN, R., Some comparison properties and bounds for capillary surfaces, Complex Analysis and its Applications, Moscow Mathem. Soc., volume dedicated to I. N. VEKUA, "Scientific Press", Moscow 1978.

Eingereicht: 16. 5. 1980

Anschrift: Professor Dr. ROBERT FINN, Standford University, Department of Mathematics, Stanford, California 94305, U.S.A.

[3]) The formal estimates of CONCUS apply (as functions of the parameter ψ) to the entire traverse of the curve.

ZAMM 61, 175–177 (1981)

R. FINN

Addenda to My Paper, "On the Laplace Formula and the Meniscus Height for a Capillary Surface" Preceding in this Issue

In two senses, the material of [1] yields significantly more information than is indicated in that reference.

1. We observe first that the method can be applied to obtain new estimates, above and below, for the height u_a of the contact line where the free surface meets the bounding wall. To be sure, some such estimates follow by adding the bounds in [1] for U_0 and for $Q = U_a - U_0$; however, the errors then also add, leading to a poor result. We proceed instead by introducing comparison circles that contact the meridional solution curve at (a, u_a). In what follows we use the notation, including numeration of formulas and figures, appearing in [1], and continue that numeration in this note.

The relation (12), together with the reasoning of section 4, shows that *a circle centered on the symmetry axis, passing through* $(0, u_0)$ *and making at* $r = a$ *the angle* ψ *with the horizontal, will lie in the range* $0 < r \leqq a$ *everywhere above the meridional section* $U(r; u_0)$. The same reasoning shows that *if the identical circle is shifted vertically downward until it meets the solution section (tangentially) at* (a, u_a), *then in the interval* $0 \leqq r < a$ *it will lie strictly below that section, and hence the corresponding rotation surface will bound, with the base plane, a smaller volume.* The formal expression of that circumstance now becomes

$$u_a < \frac{2}{\varkappa a} \sin \psi + \frac{a}{\sin \psi} \left[\frac{1}{1 + \cos \psi} - \frac{1 - \cos^3 \psi}{3 \sin^2 \psi} \right]. \tag{49}$$

The relation (49) can be viewed as a counterpart, for the edge height u_a, of the relation (16) for u_0. The result should be compared also with the estimate (25), which it improves in some cases. The latter relation, in conjunction with the estimate for u_0^+ from (18), is asymptotically the stronger one, as $B \to 0$. However, (49) is certainly simpler to apply, and it yields an improved, or at most marginally inferior result, in cases of general interest.

To obtain a lower bound for u_a, we return to the decomposition

$$k_l + k_m = \varkappa u$$

for the sectional curvatures k_l, k_m. According to the known theorem of LIOUVILLE, we have

$$k_l(a) = \frac{\sin \psi}{a}$$

and thus

$$k_m(a) = \varkappa u_a - \frac{\sin \psi}{a}.$$

Since $k_m(r)$ is increasing, it follows that a lower circular arc of radius

$$\varrho_a = \frac{a}{\varkappa a u_a - \sin \psi}$$

tangent to the meridional solution curve at (a, u_a), will lie entirely above that curve. We obtain an upper comparison surface by continuing the circular arc to its lowest point, then extending it to the axis as a straight segment (Figure 5). The comparison of volumes now yields

$$a^2 u_a - (a - \varrho_a)^2 \varrho_a (1 - \cos \psi) - (a - \varrho_a) \varrho_a^2 (\psi - \sin \psi \cos \psi) - \varrho_a^3 \frac{(1 - \cos \psi)^2}{3} (2 + \cos \psi) > 2 \frac{a}{\varkappa} \sin \psi. \tag{50}$$

Geometric considerations show that the left side of (50) is increasing in u_a, at least in any interval for which $\varkappa a u_a > 2 \sin \psi$, which by (6b) must hold on any solution curve. Thus, a lower bound for u_a is provided by the (unique) solution u_a^- of

$$\left(u - \frac{2}{\varkappa a} \sin \psi\right) (\varkappa a u - \sin \psi)^3 - (\varkappa a u - \sin \psi - 1)^2 (1 - \cos \psi)\, a -$$
$$- (\varkappa a u - \sin \psi - 1) (\psi - \sin \psi \cos \psi)\, a - \frac{(1 - \cos \psi)^2}{3} (2 + \cos \psi)\, a = 0 \tag{51}$$

for which $\varkappa a u > 2 \sin \psi$.

Again, (51) yields a result that is in some, but not in all cases, preferable to the lower bound given by (23) in conjunction with (16) or (for larger B) with (48).

2. The remark in section 12 of [1], as to the prospect for extending the results throughout the "intermediate" range, was superfluous, as would have been apparent had we made some further calculations. In fact, the maximum percentage discrepancies between upper and lower bounds for U_a and for Q appear to occur at about $B = 10$, the accuracy improving rapidly for larger B. Further, the estimates for U_a remain surprisingly (to this author) accurate for all configurations.

In Figures 4(A) and 6 are shown the results of calculations of U_a and of Q up to the value $B = 100$, which is the smallest B for which the "large BOND number" calculations of CONCUS can be considered reliable (see the remarks in reference 5, p. 492, of [1]). Table 3 gives calculated bounds for U_a, according to the varying methods. It should be noted that relative accuracy for values of U_0 at large B is of little interest for the present study, as U_0 becomes rapidly negligible with increasing B.

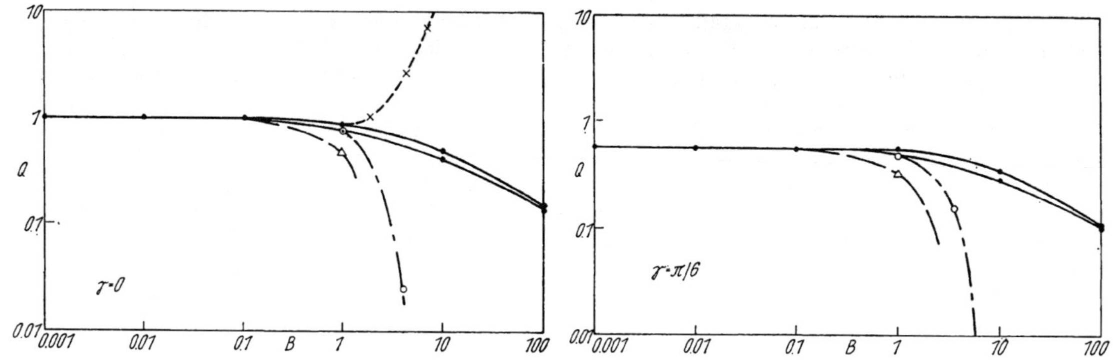

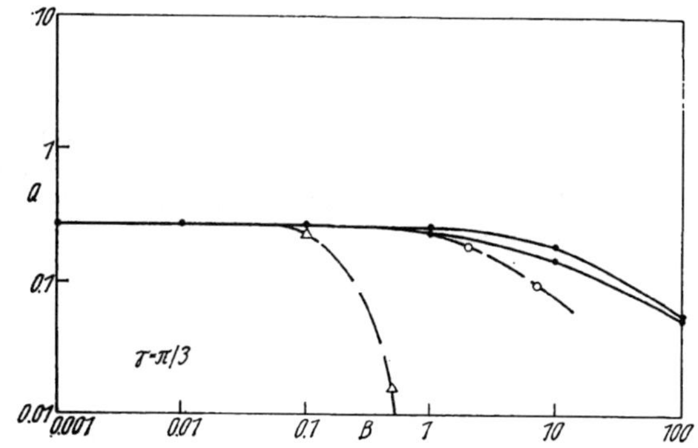

Fig. 4 (A). Meniscus heights for varying BOND number: ● Upper and lower bounds (solid lines); △ LAPLACE; ○ POISSON, CONCUS; × RAYLEIGH

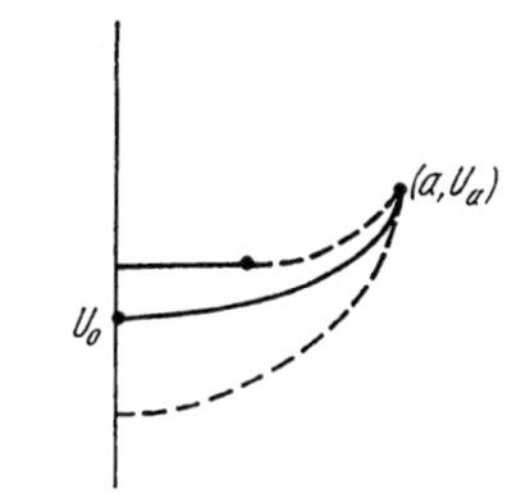

Fig. 5. Comparison circles

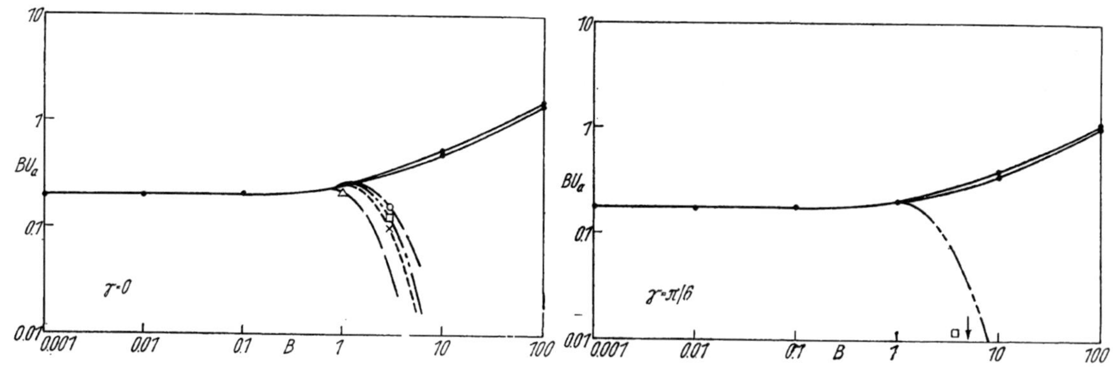

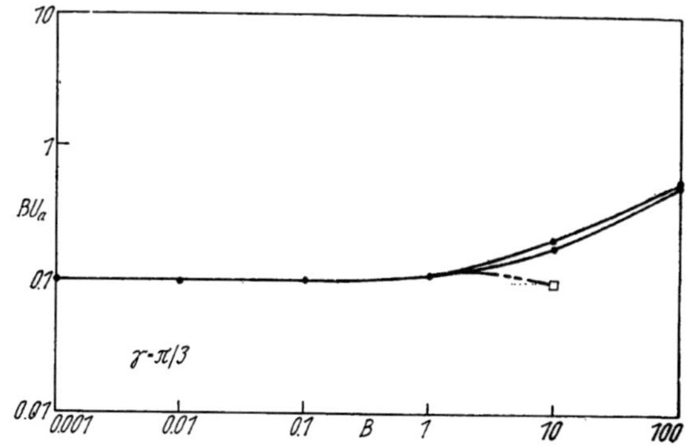

Fig. 6. Height of contact line for varying BOND number: ● Upper and lower bounds (solid lines); △ LAPLACE; ○ POISSON; × RAYLEIGH; — — CONCUS

3. We mention finally a new contribution by BRULOIS [2], who obtains, for any configuration, a sequence of upper bounds for U_0 that converges to the exact value. BRULOIS' procedure relies heavily on the analyticity of the solutions, and yields some remarkable properties of their power expansions. When adapted to the case of small B, the initial step yields precisely the upper bound determined by (18).

From the point of view of the method we have presented, a natural next step would be to extend it to a convergent iterative procedure, perhaps by determining an appropriate sequence of comparison surfaces. Such a procedure does not follow immediately from what we have presented, however we hope to return to that matter at a later time.

Table 3. Upper and lower bounds for $BU_a = BU_a/a$

$\gamma\downarrow$ $B\rightarrow$	.001	.01	.1	1	10	100
0	2.0006666	2.006663	2.06607	2.5862	5.402	15.107
	2.0006664	2.006639	2.06403	2.5034	4.963*	14.632*
$\pi/6$	1.7323716	1.735260	1.76427	2.0504	3.954	10.828
		58♯	13♯			
	3714*	5248*	6311*	827*		
	1.7322585	1.732260	1.75287	1.9391	3.581*	10.422*
$\pi/3$	1.0001372	1.001373	1.01380	1.1403	2.090	5.652
		2♯	72♯	372♯		
	1371*	1368*	1336*	1113*		
	1.000043	1.000426	1.00404	1.0286	1.878*	5.420*

♯ Computed from (49); * Computed from (23); Otherwise computed from (25) or from (51)

References

1 FINN, R., On the LAPLACE formula and the meniscus height for a capillary surface, preceding in this issue.
2 BRULOIS, F., Dissertation, Stanford University, in preparation.

Eingereicht: 18. 8. 1980

Anschrift: Professor Dr. ROBERT FINN, Stanford University, Department of Mathematics, Stanford, California 94305, U.S.A.

ZAMM 61, 179 – 191 (1981)

H. Hilbig

Existenzsatz für Potentialströmungen mit Totwassern um zwei geradlinige Hindernisse

Es werden ideale, inkompressible, wirbelfreie, ebene Strömungen um zwei geradlinige Hindernisse, hinter denen sich bis ins Unendliche erstreckende Totwassergebiete ausbilden, betrachtet. Zum Nachweis der Existenz derartiger Strömungen um Hindernisse mit vorgegebenen Längen und vorgegebenem vertikalen Abstand werden Fixpunktgleichungen aufgestellt. Ihre Lösbarkeit folgt mit Hilfe der Theorie des Abbildungsgrades, da in einem Symmetriefall eine Lösung angegeben werden kann.

Consider ideal incompressible irrotational plane flow around two rectiline obstacles. The cavities formed behind the obstacles shall be infinite in extent. The existence of such flows around obstacles with given length and given vertical distance is equivalent to the existence of a solution of a functional equation, the existence of that solution follows from the fix point theory of Leray-Schauder.

Рассматриваются безвихревые плоские течения идеальной несжимаемой жидкости вокруг двух прямолинейных препятствий, вслед которых образуются безконечные застойные области. Для доказательства существования таких течений вокруг препятствий с заданными длинами и заданным вертикальным расстоянием устанавливаются уравнения неподвижной точки. Решимость этих уравнений следует из теории степени отображения, поскольку имеется решение для симметричного случая.

1.

In einer idealen, inkompressiblen, wirbelfreien, ebenen Flüssigkeitsströmung befinden sich zwei Hindernisse. Die Flüssigkeit trifft in Staupunkten O bzw. O' auf die Hindernisse, umströmt sie und bildet hinter ihnen Totwassergebiete, in denen die Flüssigkeit ruht. Die Totwasser sollen sich bis ins Unendliche erstrecken. Sie werden vom Strömungsgebiet durch Linien abgegrenzt, die sich aus den beiden durch die Staupunkte O bzw. O' getrennten Teilen ω_1, ω_2 bzw. ω_1', ω_2' der Kontur der Hindernisse und den freien Linien λ_1, λ_2 bzw. λ_1', λ_2' zusammensetzen (Abb. 1).

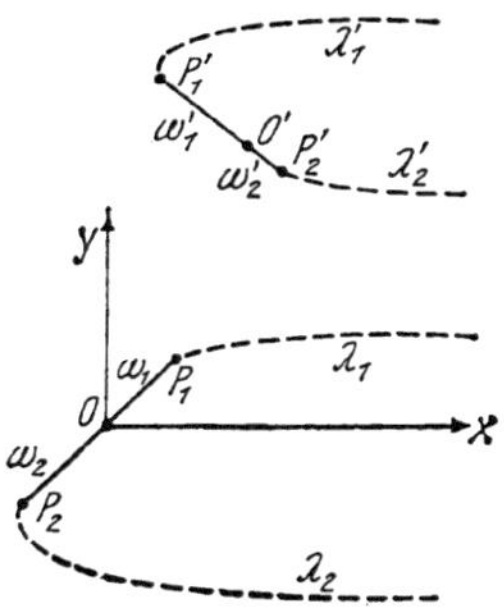

Abb. 1

Der Staupunkt O sei der Ursprung eines rechtwinkligen Koordinatensystems, dessen x-Achse in Richtung der Strömung im Unendlichen liege. Sind u und v die Komponenten der Geschwindigkeit mit dem Betrag $V = 1$ im Unendlichen, so gilt bis auf die Totwasser

$$u_\infty = 1\,, \qquad v_\infty = 0\,. \tag{1.1}$$

Die Totwasser sind Gebiete konstanten Druckes p_0. Ist die Dichte der Flüssigkeit 1, so folgt aus der Bernoullischen Gleichung $p = -\frac{1}{2}v^2 + \text{const.}$ und der Stetigkeit des Druckes, daß auf den freien Linien V konstant und wegen (1.1) $V = 1$ gilt.

Für Potentialströmungen existiert ein Geschwindigkeitspotential φ und eine Stromfunktion ψ mit den Beziehungen

$$u = \frac{\partial\varphi}{\partial x} = \frac{\partial\psi}{\partial y}\,, \qquad v = \frac{\partial\varphi}{\partial y} = -\frac{\partial\psi}{\partial x}\,.$$

Sie sind bis auf Konstanten bestimmt, die so gewählt seien, daß im Staupunkt O beide verschwinden. Die Begrenzungen der Totwasser sind Stromlinien, auf ihnen gilt daher

$$\psi = 0 \quad \text{bzw.} \quad \psi = \psi_1\,.$$

Das komplexe Potential $f = \varphi + i\psi$ ist eine im Strömungsgebiet reguläre Funktion der Variablen $z = x + iy$; $\dfrac{\mathrm{d}f}{\mathrm{d}z} = w = u - iv$ stellt den an der reellen Achse gespiegelten Geschwindigkeitsvektor dar. Ist s die Bogenlänge einer der von O bzw. O' ausgehenden Begrenzungslinien der Totwasser, so gilt wegen $\frac{dx}{ds} = \frac{u}{V}, \frac{dy}{ds} = \frac{v}{V}$ auf ihnen

$$\frac{d\varphi}{ds} = \frac{\partial\varphi}{\partial x}\frac{dx}{ds} + \frac{\partial\varphi}{\partial y}\frac{dy}{ds} = V > 0\,,$$

so daß die Funktion φ auf ihnen monoton von 0 bzw. von einem Wert φ_1 bis ins Unendliche wächst. Dem Strömungsgebiet entspricht in der f-Ebene des komplexen Potentials das in Abb. 2 dargestellte Gebiet D; dabei sind f_1, f_2, f_1', f_2' die Werte des Geschwindigkeitspotentials in den Ablösungspunkten der freien Linien von den Hindernissen.

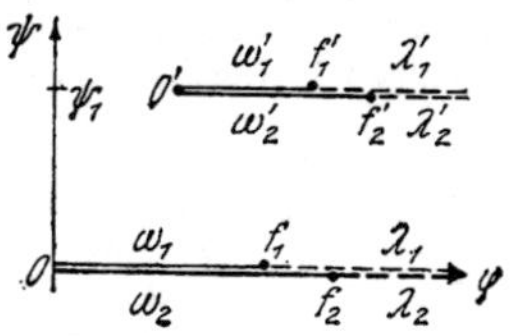

Abb. 2

Es kann das folgende Problem (Totwasserproblem) gestellt werden: Es ist eine konforme Abbildung des Gebietes D auf ein Gebiet der z-Ebene zu bestimmen, dessen Grenzen von zwei vorgegebenen Kurven mit den vorgegebenen Längen l bzw. l' und einem vorgegebenen vertikalen Abstand k der beiden Punkte P_1 und P_2' und von unbekannten freien Linien begrenzt wird. Die freien Linien sind Stromlinien und verbinden die äußersten Punkte der vorgegebenen Geraden mit dem Unendlichen, auf ihnen gilt $|dz/df| = 1$. Die Abbildung läßt den unendlich fernen Punkt und die Richtung in diesem Punkt fest.

Das Problem soll in dem Spezialfall untersucht werden, daß die beiden Hindernisse Geradenstücke sind. Zum Beweis der Existenz einer Lösung des Problems wird das Gebiet D konform auf einen Halbkreisring einer Z-Ebene abgebildet. Es können dann Fixpunktgleichungen für gewisse Parameter aufgestellt und a-priori-Schranken nachgewiesen werden. Die Existenz einer Lösung folgt mit Hilfe der Theorie des Abbildungsgrades daraus, daß in einem Symmetriefall eine Lösung vorhanden ist.

2.

Durch die Funktion $f(F)$ mit

$$\frac{df}{dF} = A\,\frac{(F - a_0)\,(F - a_0')}{F - b}$$

wird das Gebiet D zunächst auf die obere F-Halbebene abgebildet; es ist $A > 0$, $d < a_0 < -1 < b < 1 < a_0' < e$ (Abb. 3),

$$f(F) = A\,[\tfrac{1}{2}F^2 + (b - a_0 - a_0')\,F + (b - a_0)\,(b - a_0')\ln(F - b)] + c + ic'\,.$$

F-Ebene

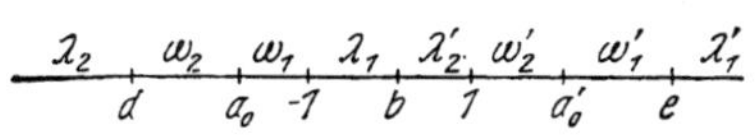

Abb. 3

Setzt man $c' = -A(b - a_0)\,(b - a_0')\,\pi$, so ist $\operatorname{Im} f = 0$ für $-\infty < F < a_0$, während $\operatorname{Re} f$ monoton von ∞ bis zu einem Minimum, das 0 gesetzt wird, fällt:

$$A[\tfrac{1}{2}a_0^2 + (b - a_0 - a_0')\,a_0 + (b - a_0)\,(b - a_0')\ln(b - a_0)] + c = 0\,.$$

Die Abbildungsfunktion lautet dann

$$f(F) = A\left[\frac{1}{2}(F^2 - a_0^2) + (b - a_0 - a_0')\,(F - a_0) + (b - a_0)\,(b - a_0')\ln\frac{F - b}{b - a_0} - i(b - a_0)\,(b - a_0')\,\pi\right]$$

$$= A\left[\frac{1}{2}(F - a_0)^2 + (b - a_0')\,(F - a_0') + (b - a_0)\,(b - a_0')\ln\frac{F - b}{b - a_0} - i(b - a_0)\,(b - a_0')\,\pi\right].$$

Für $F = d$ wird dabei der Wert f_2 erreicht. Im Intervall $a_0 < F < b$ ist $\operatorname{Im} f = 0$, während $\operatorname{Re} f$ von 0 bis ∞ monoton wächst, wobei sich für $F = -1$ der Wert f_1 ergibt. Im Intervall $b < F < a_0'$ fällt $\operatorname{Re} f$ monoton von ∞ bis zu einem Minimum, das φ_1 gesetzt wird, während für $\operatorname{Im} f = c'$ gilt. Für $F = 1$ wird dabei der Wert f_2' erreicht. Schließlich wächst $\operatorname{Re} f$ im Intervall $a_0' < F < \infty$ monoton von φ_1 bis ∞, wobei sich für $F = e$ der Wert f_1' ergibt. Daraus folgen die 6 Gleichungen

$$f_2 = \frac{1}{2} A(d - a_0)^2 + A(b - a_0')\,(d - a_0) + A(b - a_0)\,(b - a_0') \ln \frac{b - d}{b - a_0},$$

$$f_1 = \frac{1}{2} A(-1 - a_0)^2 + A(b - a_0')\,(-1 - a_0) + A(b - a_0)\,(b - a_0') \ln \frac{b + 1}{b - a_0},$$

$$f_2' = \frac{1}{2} A(1 - a_0)^2 + A(b - a_0')\,(1 - a_0) + A(b - a_0)\,(b - a_0') \ln \frac{1 - b}{b - a_0},$$

$$f_1' = \frac{1}{2} A(e - a_0)^2 + A(b - a_0')\,(e - a_0) + A(b - a_0)\,(b - a_0') \ln \frac{e - b}{b - a_0},$$

$$\varphi_1 = \frac{1}{2} A(a_0' - a_0)^2 + A(b - a_0')\,(a_0' - a_0) + A(b - a_0)\,(b - a_0') \ln \frac{a_0' - b}{b - a_0},$$

$$\psi_1 = -A(b - a_0)\,(b - a_0')\,\pi\,,$$

durch die man die Parameter A, d, a_0, b, a_0', e aus den Werten $f_1, f_2, f_1', f_2', \varphi_1, \psi_1$ berechnen kann. Setzt man nämlich

$$X = \frac{b+1}{b - a_0}, \qquad Y = \frac{b - d}{b - a_0}, \qquad U = \frac{1 - b}{b - a_0}, \qquad Z = \frac{a_0' - b}{b - a_0}, \qquad V = \frac{e - b}{b - a_0},$$

wobei $0 < X < 1$, $Y > 1$, $0 < U < Z < V$ gilt, so erhält man die Gleichungen

$$f_1 = \frac{\psi_1}{\pi}\left[\frac{(1 - X)^2}{2Z} - (1 - X) - \ln X\right], \qquad f_2 = \frac{\psi_1}{\pi}\left[\frac{(1 - Y)^2}{2Z} - (1 - Y) - \ln Y\right],$$

$$f_1' = \frac{\psi_1}{\pi}\left[\frac{(1 + V)^2}{2Z} - (1 + V) - \ln V\right], \qquad f_2' = \frac{\psi_1}{\pi}\left[\frac{(1 + U)^2}{2Z} - (1 + U) - \ln U\right],$$

$$\varphi_1 = \frac{\psi_1}{\pi}\left[\frac{(1 + Z)^2}{2Z} - (1 + Z) - \ln Z\right].$$

Die letzte Gleichung liefert zunächst genau ein Z und die übrigen Gleichungen dann X, Y, U, V mit $0 < X < 1$, $Y > 1$, $U < Z < V$. Daraus ergeben sich schließlich Werte für d, a_0, b_0 a_0', e, A mit $d < a_0 < 1 < b < a_0' < e$, $A > 0$.

Setzt man nun

$$u - \frac{1}{2}\gamma = \int\limits_{-\infty}^{F} \frac{\mathrm{d}F}{\sqrt{(F - d)\,(F^2 - 1)\,(F - e)}}$$

mit der Umkehrung

$$F = \frac{1}{2}\,\frac{\wp'\left(u - \frac{\gamma}{2}\right) - \wp'(\gamma)}{\wp\left(u - \frac{\gamma}{2}\right) - \wp(\gamma)} + \frac{d + e}{4},$$

so erhält man eine Abbildung der F-Halbebene auf ein Rechteck der u-Ebene, $2\dot\omega$ und $2\dot\omega'$ seien die Perioden der Weierstraßschen $\wp$-Funktion. Durch die Transformationen

$$v = iu - i\dot\omega'\,, \qquad v = \omega + \omega' - \frac{\omega}{i\pi} \ln Z\,, \qquad \omega = -i\dot\omega'\,, \qquad \omega' = i\dot\omega$$

ergibt sich in der Z-Ebene ein Halbkreisring (Abb. 4). Dabei entsprechen den beiden Hindernissen die Halbkreise mit den Radien 1 bzw. $q = e^{i\pi\tau}$, $\tau = \dfrac{\omega'}{\omega}$, während die freien Linien in die Segmente $-1 < Z < -q$ bzw. $q < Z < 1$ übergehen. Die Punkte e^{is_0}, $qe^{is_0'}$ sind die Bilder der beiden Staupunkte O, O', die Punkte X_1, X_2 die Bilder der im Unendlichen liegenden Schnittpunkte der freien Linien λ_1, λ_2' bzw. λ_1', λ_2.

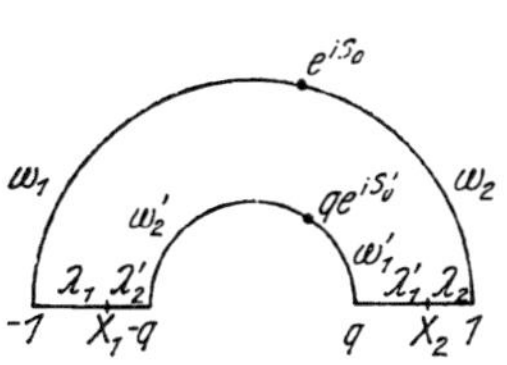

Abb. 4

Durch Zusammensetzen der durchgeführten Abbildungen ergibt sich

$$\frac{\mathrm{d}f}{\mathrm{d}Z} = \frac{\omega}{i\pi} A \frac{\wp'^2\left(\omega' - \frac{\omega}{i\pi}\ln X_2\right)}{\wp\left(\omega' - \frac{\omega}{\pi}s_0\right) - \wp\left(\omega' - \frac{\omega}{i\pi}\ln X_2\right)} \cdot \frac{\wp\left(\omega' - \frac{\omega}{i\pi}\ln X_1\right) - \wp\left(\omega' - \frac{\omega}{i\pi}\ln X_2\right)}{\wp\left(\frac{\omega}{\pi}s_0'\right) - \wp\left(\omega' - \frac{\omega}{i\pi}\ln X_2\right)} \times$$

$$\times \frac{\wp\left(\omega' - \frac{\omega}{i\pi}\ln Z\right) - \wp\left(\omega' - \frac{\omega}{\pi}s_0\right)}{\left[\wp\left(\omega' - \frac{\omega}{i\pi}\ln Z\right) - \wp\left(\omega' - \frac{\omega}{i\pi}\ln X_2\right)\right]^3} \cdot \frac{\wp\left(\omega' - \frac{\omega}{i\pi}\ln Z\right) - \wp\left(\frac{\omega}{\pi}s_0'\right)}{\wp\left(\omega' - \frac{\omega}{i\pi}\ln Z\right) - \wp\left(\omega' - \frac{\omega}{i\pi}\ln X_1\right)} \cdot \frac{\wp'\left(\omega' - \frac{\omega}{i\pi}\ln Z\right)}{Z}$$

$$= A f_1(Z) \frac{1}{iZ}.$$

Für den Fluß ψ_1 zwischen λ_1 und λ_2' folgt

$$\psi_1 = A\pi \frac{\wp'^2\left(\omega' - \frac{\omega}{i\pi}\ln X_2\right)}{\left[\wp\left(\omega' - \frac{\omega}{\pi}s_0\right) - \wp\left(\omega' - \frac{\omega}{i\pi}\ln X_2\right)\right]\left[\wp\left(\frac{\omega}{\pi}s_0'\right) - \wp\left(\omega' - \frac{\omega}{i\pi}\ln X_2\right)\right]} \times$$

$$\times \frac{\wp\left(\omega' - \frac{\omega}{i\pi}\ln X_1\right) - \wp\left(\omega' - \frac{\omega}{\pi}s_0\right)}{\wp\left(\omega' - \frac{\omega}{i\pi}\ln X_1\right) - \wp\left(\omega' - \frac{\omega}{i\pi}\ln X_2\right)} \cdot \frac{\wp\left(\omega' - \frac{\omega}{i\pi}\ln X_1\right) - \wp\left(\frac{\omega}{\pi}s_0'\right)}{\wp\left(\omega' - \frac{\omega}{i\pi}\ln X_1\right) - \wp\left(\omega' - \frac{\omega}{i\pi}\ln X_2\right)}.$$

3.

Die komplexe Geschwindigkeit w ist ebenso wie $f(z)$ eine reguläre Funktion im Strömungsgebiet. Durch Einführung der Umkehrfunktion von $f(z)$ wird sie eine reguläre Funktion im Gebiet D bzw. im Halbkreisring der Z-Ebene.

Wird durch $w = e^{i\omega}$ die Funktion $\omega(Z) = \vartheta(Z) + i\tau(Z)$ eingeführt, so gilt $u = e^{\tau}\cos\vartheta$, $v = e^{\tau}\sin\vartheta$, $V = e^{\tau}$, und $\omega(Z)$ ist eine reguläre Funktion im Halbkreisring. Der Realteil $\vartheta(Z)$ von $\omega(Z)$ stellt den Winkel der Geschwindigkeit in dem Z entsprechenden Punkt der Strömungsebene mit der reellen Achse dar. Da die Flüssigkeit im Unendlichen parallel zur reellen Achse strömt, gilt $\vartheta(X_1) = \vartheta(X_2) = 0$. Wegen $V = 1$ auf den freien Linien ist $\omega(Z)$ auf der reellen Achse reell, sie kann daher durch Spiegelung in die untere Z-Halbebene fortgesetzt werden, und man erhält eine im Ringgebiet $q < |Z| < 1$ reguläre Funktion.

4.

Wird eine in einem Halbkreisring $q < |Z| < 1$ reguläre Funktion $\omega(Z)$ vorgegeben, die auf der reellen Achse reelle Werte mit $\vartheta(X_1) = \vartheta(X_2) = 0$ für $-1 < X_1 < -q$, $q < X_2 < 1$ besitzt, so liefert die Funktion $\mathrm{d}z = e^{i\omega}\,\mathrm{d}f$ eine konforme Abbildung auf ein Gebiet der z-Ebene. Die Bilder der Segmente $-1 < Z < -q$, $q < Z < 1$ sind Kurven, die sich ins Unendliche erstrecken, dort wegen $\vartheta(X_1) = \vartheta(X_2) = 0$ parallel zur x-Achse liegen und die Bedingung $|\mathrm{d}z| = |\mathrm{d}f|$ erfüllen. Diese Abbildung ist die Lösung des Totwasserproblems, wenn die Halbkreise $|Z| = 1$, $Y > 0$ bzw. $|Z| = q$, $Y > 0$ auf vorgegebene Hindernisse mit Längen l bzw. l' abgebildet werden und der vertikale Abstand der Ablösungspunkte P_1 und P_2' einen vorgegebenen Wert k hat. Ferner dürfen sich die freien Linien hinter den Hindernissen nicht schneiden. Die beiden Hindernisse seien durch ihre Winkel $n\pi$ bzw. $m\pi$ mit der x-Achse vorgegeben ($0 < n < 1$, $0 < m < 1$); als Winkel der Strömungsgeschwindigkeit ergibt sich daraus

$$\vartheta(e^{is}) = \begin{cases}(n-1)\pi & \text{für } 0 < s < s_0\\ n\pi & \text{für } s_0 < s < \pi\end{cases}, \qquad \vartheta(qe^{is}) = \begin{cases}m\pi & \text{für } 0 < s < s_0'\\ (m-1)\pi & \text{für } s_0' < s < \pi\end{cases}. \tag{4.1}$$

Die für $q < |Z| < 1$ reguläre Funktion, deren Realteile auf den Kreisen $|Z| = 1$ bzw. $|Z| = q$ vorgegebene Werte $\varphi(s)$ bzw. $\psi(s)$ besitzt, wird durch die Formel

$$R(z) = \frac{i\omega}{\pi^2}\int_0^{2\pi}\varphi(s)\left\{\frac{\sigma'\left(\frac{\omega}{i\pi}\ln Z - \frac{\omega}{\pi}s\right)}{\sigma\left(\frac{\omega}{i\pi}\ln Z - \frac{\omega}{\pi}s\right)} + \left(\frac{1}{2\omega'} - \frac{\eta}{i\pi}\ln Z\right)\right\}\mathrm{d}s -$$

$$- \frac{i\omega}{\pi^2}\int_0^{2\pi}\psi(s)\left\{\frac{\sigma_3'\left(\frac{\omega}{i\pi}\ln Z - \frac{\omega}{\pi}s\right)}{\sigma_3\left(\frac{\omega}{i\pi}\ln Z - \frac{\omega}{\pi}s\right)} + \left(\frac{1}{2\omega'} - \frac{\eta}{i\pi}\ln Z\right)\right\}\mathrm{d}s$$

geliefert, sie ist eindeutig im Kreisring, falls

$$\int_0^{2\pi} [\varphi(s) - \psi(s)]\, ds = 0$$

gilt. Diese Bedingung ist bei obigen Randwerten für $s_0 + s_0' = \pi - \alpha, \pi - \alpha = (m - n)\pi$ erfüllt. Die Parameter s_0, s_0' können daher für $\alpha > 0$ bzw. $\alpha < 0$ nur in den Intervallen $0 < s_0 < \pi - \alpha$, $\pi - \alpha > s_0' > 0$ bzw. $-\alpha < s_0 < \pi$, $\pi > s_0' > -\alpha$ variieren. Berechnet man die Funktion $R(Z)$ mit den Randwerten (4.1), so folgt

$$\omega(Z) = i \ln \frac{\vartheta_1(u_0 - v)\, \vartheta_0(u_0' - v)}{\vartheta_1(u_0 + v)\, \vartheta_0(u_0' + v)} + (n - 1)\pi \tag{4.2}$$

mit $u_0 = \frac{1}{2\pi} s_0$, $u_0' = \frac{1}{2\pi} s_0'$, $v = \frac{1}{2\pi i} \ln Z$.

5.

Die Richtungen der freien Linien werden durch die Werte der auf den Segmenten $-1 < X < -q$, $q < X < 1$ reellen Funktion $\omega(Z)$ bestimmt. Da sie in den Endpunkten dieser Segmente unterschiedliche Vorzeichen besitzt, existiert mindestens eine Nullstelle X_1 in $-1 < X < -q$ und mindestens eine Nullstelle X_2 in $q < X < 1$ von $\omega(Z)$. Sind dies die einzigen Nullstellen, dann ist

$$\vartheta(X) > 0 \quad \text{in} \quad -1 < X < X_1, \qquad \vartheta(X) < 0 \quad \text{in} \quad X_2 < X < 1,$$
$$\vartheta(X) < 0 \quad \text{in} \quad X_1 < X < -q, \qquad \vartheta(X) < 0 \quad \text{in} \quad q < X < X_2,$$

und die freien Linien können sich hinter den Hindernissen nicht schneiden. Um dies festzustellen, wird das Monotonieverhalten von $\vartheta(X)$ durch Differentiation untersucht. In $q < X < 1$ gilt $\frac{i\tau}{2} < v < 0$, und für die Funktion $L_1(v) = \frac{1}{i} \ln \frac{\vartheta_1(u_0 + iv)}{\vartheta_1(u_0 - iv)}$ folgt wegen

$$\frac{\vartheta_1'(v)}{\vartheta_1(v)} = 2\omega\zeta(v_1) - 2\eta v_1$$

$$\frac{\partial L_1}{\partial v} = 2\omega\, \{\zeta(u_1 + iv_1) + \zeta(u_1 - iv_1)\} - 4\eta u_1$$

mit $v_1 = 2\omega v$, $u_1 = 2\omega u_0$. Weiter gilt

$$\frac{\partial^2 L_1}{\partial v^2} = (2\omega)^2 \frac{\wp'(u_1)\, i\wp'(iv_1)}{[\wp(u_1) - \wp(iv_1)]^2} > 0,$$

da $\wp'(u_1) < 0$ für $0 < u_1 < \omega$, $i\wp'(iv_1) < 0$ für $\frac{i\tau}{2} < v < 0$ ist. Daher ist $\frac{\partial L_1}{\partial v}$ in $\frac{i\tau}{2} < v < 0$ monoton wachsend:

$$\frac{\partial L_1}{\partial v}(v) > \frac{\partial L_1}{\partial v}\left(\frac{i\tau}{2}\right) = 2\omega\{\zeta(u_1 - \omega') + \zeta(u_1 + \omega')\} - 4\eta u_1 = M(u_1).$$

Es ist $M'(u_1) = -4\omega\wp(u_1 + \omega') - 4\eta$ monoton fallend mit $M(0) = M(\omega) = 0$; daher gilt $\frac{\partial L}{\partial v}\left(\frac{i\tau}{2}\right) = M(u_1) \geqq 0$, und $L_1(v)$ ist monoton wachsend. Auf die gleiche Weise folgt das monotone Wachsen der Funktion

$$L_2(v) = \frac{1}{i} \ln \frac{\vartheta_0(u_0' + iv)}{\vartheta_0(u_0 - iv)},$$

so daß $\vartheta(X) = -L_1(v) - L_2(v)$ in $q < X < 1$ monoton fällt. Ebenso zeigt man die Monotonie von $\vartheta(X)$ in $-1 < < X < -q$.

6.

Durch Vorgabe von $q(0 < q < 1)$ und $s_0(0 < s_0 < \pi - \alpha$, bzw. $-\alpha < s_0 < \pi)$ ist die Funktion $\omega(Z)$ festgelegt, sie besitzt in $-1 < X < -q$ bzw. $q < X < 1$ genau eine Nullstelle. Wegen $\frac{df}{dz} = w = e^{-i\omega}$ wird durch $dz = e^{i\omega}\, df$ der Halbring auf ein Gebiet der z-Ebene abgebildet:

$$z = A \int_{e^{is_0}}^{Z} e^{i\omega(\zeta)} f_1(\zeta) \frac{d\zeta}{i\zeta}.$$

Den Punkten $Z = e^{is}$ bzw. $Z = q\, e^{is}$ entsprechen dabei geradlinige Stücken mit den Winkeln $n\pi$ bzw. $m\pi$ zur x-Achse und den Bogenlängen $L(s)$ bzw. $L'(s)$, die sich aus

$$L(s) = A \int_0^s e^{-\tau(e^{is})}\, |f_1(e^{is})|\, ds, \qquad L'(s) = A \int_0^s e^{-\tau(qe^{is})}\, |f_1(q)\, e^{is}|\, ds$$

ergeben. Durch Wahl von A kann erreicht werden, daß das untere Hindernis eine vorgegebene Länge l erhält:

$$l = A \int_0^\pi e^{-\tau(e^{is})} |f_1(e^{is})| \, ds \,. \tag{6.1}$$

Der vertikale Abstand der beiden Ablösungspunkte P_1 und P_2' beträgt

$$K(q, s_0) = A \int_{-1}^{-q} \sin \vartheta(\xi) f_1(\xi) \frac{d\xi}{i\xi} + \psi_1 \,. \tag{6.2}$$

In der Abbildungsfunktion sind noch die beiden Parameter s_0 und q frei wählbar. Sie sollen dadurch festgelegt werden, daß das obere Hindernis die vorgegebene Länge l' und der vertikale Abstand von P_1 und P_2' den vorgegebenen Wert k erhält. Es sind also die Gleichungen

$$L'(q, s_0) \equiv A \int_0^\pi e^{-\tau(qe^{is})} |f_1(qe^{is})| \, ds = l' \,, \qquad K(q, s_0) \equiv A \int_{-1}^{-q} \sin \vartheta(\xi) f_1(\xi) \frac{d\xi}{i\xi} + \psi_1 = k \tag{6.3), (6.4}$$

zu erfüllen, die auch in der Form

$$s_0 = s_0 + L'(q, s_0) - l' \equiv \mathcal{L}(q, s_0, l') \,, \qquad q = q + K(q, s_0) - k \equiv \mathcal{K}(q, s_0, k) \tag{6.5), (6.6}$$

von Fixpunktgleichungen geschrieben werden können.

7.

Lemma: Es sei $n = m = \frac{1}{2}$; dann gilt $L'(q, s_0) = l$ genau für $s_0 = s_0'$.

Beweis: Es sei $\omega_{n,m}(Z, s_0, s_0')$ die Funktion (4.2) mit den Randwerten $n\pi$ bzw. $m\pi$ auf den Kreisen $|Z| = 1$ bzw. $|Z| = q$ zu den Parameterwerten s_0 und s_0'. Setzt man $Z = \dfrac{q}{Z'}$, (Spiegelung am Kreis $|Z| = \sqrt{q}$) und benutzt die Eigenschaften der Thetafunktionen bei Änderung des Arguments um $\dfrac{\tau}{2}$, so ergibt sich wegen $u_0 + u_0' = \dfrac{1}{2\pi}(\pi - \alpha)$

$$\omega_{n,m}(Z, s_0, s_0') = -i \ln \frac{\vartheta_1(u_0' + v)\, \vartheta_0(u_0 + v)}{\vartheta_1(u_0' - v)\, \vartheta_0(u_0 - v)} + m\pi = -\omega_{1-m,1-n}(Z', s_0', s_0) \,.$$

Wenn daher $n + m = 1$ gilt und X_1, X_2 die Nullstellen von $\omega_{n,m}(Z, s_0, s_0')$ auf der reellen Achse sind, so sind $X_1' = \dfrac{q}{X_1}$, $X_2' = \dfrac{q}{X_2}$ die Nullstellen von $\omega_{n,m}(Z, s_0', s_0)$. Für $s_0 = s_0' \left(= \dfrac{\pi - \alpha}{2}\right)$ müssen sie übereinstimmen, und man hat also $X_1 = -\sqrt{q}$, $X_2 = +\sqrt{q}$ für diesen Fall. In den Punkten $Z = e^{is}$ bzw. $Z = q\, e^{is}$ gilt

$$\begin{aligned} \tau_{n,m}(e^{is}, s_0, s_0') &= -\tau_{n,m}(q\, e^{is}, s_0', s_0) = \tau_{n,m}(q\, e^{is}, s_0', s_0) \,, \\ \tau_{n,m}(q\, e^{is}, s_0, s_0') &= -\tau_{n,m}(e^{-is}, s_0', s_0) = \tau_{n,m}(e^{is}, s_0', s_0) \,. \end{aligned} \tag{7.1}$$

Wegen

$$\wp\left(\omega' - \frac{\omega}{i\pi} \ln Z_1\right) - \wp\left(\omega' - \frac{\omega}{i\pi} \ln Z_2\right) = \frac{(e_3 - e_1)(e_3 - e_2)}{\left[\wp\left(\frac{\omega}{i\pi} \ln Z_1\right) - e_3\right]\left[\wp\left(\frac{\omega}{i\pi} \ln Z_2\right) - e_3\right]} \left[\wp\left(\frac{\omega}{i\pi} \ln Z_1\right) - \wp\left(\frac{\omega}{i\pi} \ln Z_2\right)\right],$$

$$\wp'\left(\omega' - \frac{\omega}{i\pi} \ln Z\right) = -\frac{(e_3 - e_1)(e_3 - e_2)}{\left[\wp\left(\frac{\omega}{i\pi} \ln Z\right) - e_3\right]^2} \wp'\left(\frac{\omega}{i\pi} \ln Z\right),$$

erhält man für die in der Abbildungsfunktion auftretende Funktion $f_1(Z)$, wenn man sie in Abhängigkeit von den Parametern s_0, s_0' und den Nullstellen X_1, X_2 betrachtet:

$$\begin{aligned} f_1(Z, s_0, s_0', X_1, X_2) = \frac{\omega}{\pi} & \frac{\wp'^2\left(\frac{\omega}{i\pi} \ln X_2\right)}{\wp\left(\frac{\omega}{\pi} s_0\right) - \wp\left(\frac{\omega}{i\pi} \ln X_2\right)} \cdot \frac{\wp\left(\frac{\omega}{i\pi} \ln X_2\right) - \wp\left(\frac{\omega}{i\pi} \ln X_1\right)}{\wp\left(\omega' - \frac{\omega}{\pi} s_0'\right) - \wp\left(\frac{\omega}{i\pi} \ln X_2\right)} \times \\ & \times \frac{\wp\left(\frac{\omega}{i\pi} \ln Z\right) - \wp\left(\omega' - \frac{\omega}{\pi} s_0'\right)}{\wp\left(\frac{\omega}{i\pi} \ln Z\right) - \wp\left(\frac{\omega}{i\pi} \ln X_1\right)} \cdot \frac{\wp\left(\frac{\omega}{i\pi} \ln Z\right) - \wp\left(\frac{\omega}{\pi} s_0\right)}{\wp\left(\frac{\omega}{i\pi} \ln Z\right) - \wp\left(\frac{\omega}{i\pi} \ln X_1\right)} \wp'\left(\frac{\omega}{i\pi} \ln Z\right). \end{aligned} \tag{7.2}$$

Führt man darin die Transformation $Z = q/Z'$ aus, so ergibt sich

$$f_1(Z, s_0, s_0', X_1, X_2) = -f_1(Z', s_0', s_0, X_1', X_2'); \tag{7.3}$$

insbesondere ist für $Z = e^{is}$

$$\begin{aligned} f_1(e^{is}, s_0, s_0', X_1, X_2) &= -f_1(q\, e^{is}, s_0', s_0, X_1', X_2') = f_1(q\, e^{is}, s_0', s_0, X_1', X_2')\,, \\ f_1(qe^{is}, s_0, s_0', X_1, X_2) &= -f_1(e^{-is}, s_0', s_0, X_1', X_2') = f_1(e^{is}, s_0', s_0, X_1', X_2')\,. \end{aligned} \tag{7.4}$$

Daraus folgt für $s_0 = s_0' = (\pi - \alpha)/2$

$$L'\left(q, \frac{\pi-\alpha}{2}\right) = A \int_0^\pi e^{-\tau(qe^{is}, s_0, s_0)}\, |f_1(qe^{is}, s_0, s_0)|\, ds = A \int_0^\pi e^{-\tau(e^{is}, s_0, s_0)}\, |f_1(e^{is}, s_0, s_0)|\, ds = l$$

wegen Gleichung (6.1).

Es sei nun speziell $n = m = \frac{1}{2}$. Dann gilt $s_0 + s_0' = \pi$. Um zu zeigen, daß

$$L'(q, s_0) < l \quad \text{für} \quad 0 < s_0 < \frac{\pi}{2}\,, \qquad L'(q, s_0) > l \quad \text{für} \quad \frac{\pi}{2} < s_0 < \pi$$

gilt, ist wegen der Gleichungen (6.1) und (6.3) und der Symmetrieeigenschaften (7.1) und (7.4) die Differenz

$$J = A \int_0^\pi \{e^{-\tau(e^{is}, s_0 s_0', X_1, X_2)}\, |f_1(e^{is}, s_0, s_0', X_1, X_2)| - e^{-\tau(e^{is}, s_0', s_0, X_1', X_2')}\, |f_1(e^{is}, s_0', s_0, X_1', X_2')|\}\, ds$$

zu untersuchen. Drückt man $f_1(Z)$ durch Thetafunktionen aus, so ergibt sich

$$\begin{aligned} f_1(Z, s_0, s_0', X_1, X_2) = {} & \left(\frac{\pi}{\omega}\right)^2 \vartheta_1^2(v_2)\, \vartheta_2^2(v_2)\, \vartheta_3^2(v_2)\, \vartheta_0^2(v_2) \times \\ & \times \frac{\vartheta_1(v_2 - v_1)\, \vartheta_1(v_2 + v_1)}{\vartheta_1(u_0 - v_2)\, \vartheta_1(u_0 + v_2)\, \vartheta_0(u_0' - v_2)\, \vartheta_0(u_0' + v_2)}\, \vartheta_1(v)\, \vartheta_2(v)\, \vartheta_3(v)\, \vartheta_0(v) \times \\ & \times \frac{\vartheta_1(u_0 - v)\, \vartheta_1(u_0 + v)\, \vartheta_0(u_0' - v)\, \vartheta_0(u_0' + v)}{\vartheta_1^3(v - v_2)\, \vartheta_1^3(v + v_2)\, \vartheta_1(v - v_1)\, \vartheta_1(v + v_1)} \end{aligned} \tag{7.5}$$

mit $v_1 = \frac{1}{2\pi i} \ln X_1$, $v_2 = \frac{1}{2\pi i} \ln X_2$. Verwendet man Periodizitätseigenschaften der Thetafunktionen, so folgt aus (4.2), daß zwischen den Nullstellen X_1, X_2 bzw. v_1, v_2 die Beziehung $v_1 + v_2 = \frac{1}{2} + \frac{\tau}{2}$ gilt. Zwischen u_0 und v_2 ergibt sich daraus ferner der Zusammenhang

$$\frac{\vartheta_1(u_0)\, \vartheta_3(u_0)}{\vartheta_2(u_0)\, \vartheta_0(u_0)} = -i\, \frac{\vartheta_1(v_2)\, \vartheta_3(v_2)}{\vartheta_2(v_2)\, \vartheta_0(v_2)}\,,$$

aus dem man erkennt, daß für $0 < s_0 < \frac{\pi}{2}$, d. h. $0 < u_0 < \frac{1}{4}$ die Nullstelle v_2 zwischen 0 und $\frac{\tau}{4}$ und für $\frac{1}{4} < u_0 < \frac{1}{2}$ zwischen $\frac{\tau}{4}$ und $\frac{\tau}{2}$ liegt.

Zusammen mit der aus (4.2) folgenden Darstellung

$$\tau(e^{is}, s_0, s_0', X_1, X_2) = \ln \left| \frac{\vartheta_1(u_0 - u)\, \vartheta_0(u_0' - u)}{\vartheta_1(u_0 + u)\, \vartheta_0(u_0' + u)} \right| \tag{7.6}$$

folgt dann

$$\begin{aligned} J = 2\pi \left(\frac{\pi}{\omega}\right)^2 A \int_0^{1/2} & \frac{\vartheta_1(u)\, \vartheta_2(u)\, \vartheta_3(u)\, \vartheta_0(u)}{|\vartheta_1^3(u + v_2)\, \vartheta_0^3(u + v_2)\, \vartheta_2(u + v_2)\, \vartheta_3(u + v_2)|} \times \\ & \times \{|\vartheta_0^3(u + v_2)\, \vartheta_2(u + v_2)|\, \vartheta_0^2(u + u_0)\, \vartheta_3^2(u - u_0) - |\vartheta_1^3(u + v_2)\, \vartheta_3(u + v_2)|\, \vartheta_2^2(u - u_0)\, \vartheta_0^2(u + u_0)\}\, du\,. \end{aligned}$$

Aus den Beziehungen

$$\vartheta_1(u - v_2)\, \vartheta_1(u + v_2) = \frac{1}{\vartheta_3^2}\, \{\vartheta_3^2(v_2)\, \vartheta_1^2(u) - \vartheta_1^2(v_2)\, \vartheta_3^2(u)\}\,,$$

$$\vartheta_0(u - v_2)\, \vartheta_0(u + v_2) = \frac{1}{\vartheta_3^2}\, \{\vartheta_0^2(v_2)\, \vartheta_1^2(u) + \vartheta_1^2(v_2)\, \vartheta_3^2(u)\}\,,$$

$$\vartheta_0^2(v_2) + \vartheta_1^2(v_2) = \frac{k'}{1-k}\, \{\vartheta_3^2(v_2) - \vartheta_2^2(v_2)\} = \frac{k'}{1-k}\, \vartheta_0\left(0, \frac{\tau}{2}\right) \vartheta_0\left(v_2, \frac{\tau}{2}\right)$$

folgt

$$|\vartheta_1^2(u+v_2)| - |\vartheta_0^2(u+v_2)| = \frac{1}{\vartheta_3^2}\left\{\vartheta_1^2(u) - \frac{k'}{1-k}\vartheta_3^2(u)\right\}\vartheta_0\left(0, \frac{\tau}{2}\right)\vartheta_2\left(v_2, \frac{\tau}{2}\right)$$

mit $\vartheta_1^2(u) - \frac{k'}{1-k}\vartheta_3^2(u) \leqq 0,$

$$\vartheta_0\left(v_2, \frac{\tau}{2}\right) > 0 \quad v_2 \text{ zwischen } 0 \text{ und } \frac{\tau}{4}, \quad \vartheta_0\left(v_2, \frac{\tau}{2}\right) < 0 \quad v_2 \text{ zwischen } \frac{\tau}{4} \text{ und } \frac{\tau}{2},$$

so daß

$$|\vartheta_0(u+v_2)|^2 > |\vartheta_1(u+v_2)|^2 \quad v_2 \text{ zwischen } 0 \text{ und } \frac{\tau}{4}, \quad |\vartheta_0(u+v_2)|^2 < |\vartheta_1(u+v_2)|^2 \quad v_2 \text{ zwischen } \frac{\tau}{4} \text{ und } \frac{\tau}{2}$$

gilt. Damit wird für v_2 zwischen 0 und $\frac{\tau}{4}$

$$J > 2\pi A\left(\frac{\pi}{\omega}\right)^2 \int_0^{1/2} \frac{\vartheta_1(u)\,\vartheta_2(u)\,\vartheta_3(u)\,\vartheta_0(u)}{|\vartheta_1(u+v_2)\,\vartheta_0^3(u+v_2)\,\vartheta_2(u+v_2)\,\vartheta_3(u+v_2)|} \times$$

$$\times \{|\vartheta_0(u+v_2)\,\vartheta_2(u+v_2)|\,\vartheta_1^2(u+u_0)\,\vartheta_3^2(u+u_0) - |\vartheta_1(u+v_2)\,\vartheta_3(u+v_2)|\,\vartheta_2^2(u-u_0)\,\vartheta_0^2(u+u_0)\}\,du\,.$$

Zieht man den ersten Faktor vor das Integral, so stellt man für den verbleibenden Integranden $\Theta(u)$ die Eigenschaft

$$\int_{1/4}^{1/2} \Theta(u)\,du = -\int_0^{1/4} \Theta(u)\,du$$

fest, so daß $J > 0$ folgt. Analog ergibt sich $J < 0$ für v_2 zwischen $\frac{\tau}{4}$ und $\frac{\tau}{2}$.

8.

Lemma: Die Funktion $K(q, s_0)$ in (6.2) erfüllt für $s_0 = s_0'$, $n + m = 1$ die Bedingungen

$$\lim_{q\to+0} K(q, s_0) = \infty\,, \qquad \lim_{q\to 1-0} K(q, s_0) = 0\,.$$

Beweis:

a) Der Fall $q \to 1 - 0$ ist wegen $q = e^{+i\pi\tau}$, $\tau = it$ durch $t \to +0$ gekennzeichnet. Aus den Beziehungen

$$\vartheta_1(w) = i\sqrt{\frac{i}{\tau}}\,e^{-\frac{i\pi}{\tau}w^2}\,\vartheta_1\left(\frac{w}{\tau}, -\frac{1}{\tau}\right),$$

$$\vartheta_2(w) = \sqrt{\frac{i}{\tau}}\,e^{-\frac{i\pi}{\tau}w^2}\,\vartheta_0\left(\frac{\omega}{\tau}, -\frac{1}{\tau}\right),$$

$$\vartheta_3(w) = \sqrt{\frac{i}{\tau}}\,e^{-\frac{i\pi}{\tau}w^2}\,\vartheta_3\left(\frac{w}{\tau}, -\frac{1}{\tau}\right),$$

$$\vartheta_0(w) = \sqrt{\frac{i}{\tau}}\,e^{-\frac{i\pi}{\tau}w^2}\,\vartheta_2\left(\frac{w}{\tau}, -\frac{1}{\tau}\right)$$

erhält man das Verhalten der Thetafunktionen für $t \to +0$:

$$\begin{aligned}
\vartheta_1(w) &= i\sqrt{\frac{i}{\tau}}\,e^{-\frac{i\pi}{\tau}\left(w-\frac{1}{2}\right)^2}\{1 + O(1)\}\,, && \operatorname{Re} w > 0\,,\\
\vartheta_1(w) &= 2i\sqrt{\frac{i}{\tau}}\,e^{-\frac{i\pi}{\tau}\left(w^2+\frac{1}{4}\right)}\sin\frac{i\pi}{\tau}w\,\{1 + O(1)\}\,, && \operatorname{Re} w = 0\,,\\
\vartheta_2(w) &= \sqrt{\frac{i}{\tau}}\,e^{-\frac{i\pi}{\tau}w^2}\{1 + O(1)\}\,, && -\frac{1}{2} < \operatorname{Re} w < \frac{1}{2}\\
\vartheta_3(w) &= \sqrt{\frac{i}{\tau}}\,e^{-\frac{i\pi}{\tau}w^2}\{1 + O(1)\}\,, && -\frac{1}{2} < \operatorname{Re} w < \frac{1}{2}\\
\vartheta_0(w) &= \sqrt{\frac{i}{\tau}}\,e^{-\frac{i\pi}{\tau}\left(w-\frac{1}{2}\right)^2}\{1 + O(1)\}\,, && \operatorname{Re} w > 0\,,\\
\vartheta_0(w) &= 2\sqrt{\frac{i}{\tau}}\,e^{-\frac{i\pi}{\tau}\left(w^2+\frac{1}{4}\right)}\cos\frac{i\pi}{\tau}w\,\{1 + O(1)\}\,, && \operatorname{Re} w = 0\,.
\end{aligned} \tag{8.1}$$

Setzt man

$$g = \vartheta_1^2(v_2)\,\vartheta_2^2(v_2)\,\vartheta_3^2(v_2)\,\vartheta_0^2(v_2)\,\frac{\vartheta_1(v_2 - v_1)\,\vartheta_1(v_2 + v_2)}{\vartheta_1(u_0 - v_2)\,\vartheta_1(u_0 + v_2)\,\vartheta_0(u_0' - v_2)\,\vartheta_0(u_0' + v_2)}\,,$$

$$g_1(v) = \vartheta_1(v)\,\vartheta_2(v)\,\vartheta_3(v)\,\vartheta_0(v)\,\frac{\vartheta_1(u_0 + v)\,\vartheta_1(u_0 - v)\,\vartheta_0(u_0' + v)\,\vartheta_0(u_0' - v)}{\vartheta_1^3(v - v_2)\,\vartheta_1^3(v + v_2)\,\vartheta_2(v - v_1')\,\vartheta_2(v + v_1')}$$

mit $v_1 = \frac{1}{2} + v_1'$, so ist $f_1(Z) = \left(\frac{\pi}{\omega}\right)^2 A g g_1(v)$ und

$$l = -2\pi A\left(\frac{\pi}{\omega}\right)^2 g\int_0^{1/2} e^{-\tau(e^{it})}\,g_1(u)\,du\,; \tag{8.2}$$

es folgt für $t \to +0$ aus (8.1)

$$l = \frac{\tau}{2\pi i(u_0 + u_0')}\,\pi A\left(\frac{\pi}{\omega}\right)^2 g\,e^{-\frac{2\pi}{\tau}\{u_0^2 + u_0'^2 - u_0 - u_0' - 3v_2^2 - v_1'^2\}}\,\{1 + O(1)\}\,. \tag{8.3}$$

Drückt man den Fluß ψ_1 nach Formel (2.2) durch Thetafunktionen aus, so wird

$$\psi_1 = \pi A\left(\frac{\pi}{\omega}\right)^2 g\,\frac{\vartheta_2(u_0 - v_1')\,\vartheta_2(u_0 + v_1')\,\vartheta_3(u_0' - v_1')\,\vartheta_3(u_0' + v_1')}{\vartheta_2^3(v_2 - v_1')\,\vartheta_2^3(v_2 + v_1')}\,. \tag{8.4}$$

Zusammen mit (8.3) erhält man daraus das Verhalten

$$\psi_1 = 2\pi(u_0 + u_0')\,e^{-\frac{2\pi i}{\tau}(u_0 + u_0')}\,l\{1 + O(1)\}$$

für $t \to 0$, so daß wegen $u_0 + u_0' = \frac{\pi - \alpha}{2\pi} > 0$ $\lim\limits_{q\to 1-0} \psi_1 = 0$ gilt.

Den ersten Summanden in (6.2) kann man wegen der Symmetrieeigenschaft (7.3) in

$$I = 2A\int_{-1}^{-\sqrt{q}} \sin\vartheta(\xi)\,f_1(\xi)\,\frac{d\xi}{i\xi}$$

umformen. Mit $v' = \frac{1}{2i\pi}\ln\xi - \frac{1}{2} = v - \frac{1}{2}$ folgt daraus

$$I = 4\pi A\int_0^{\tau/4} \sin\vartheta(v')\,f_1(v')\,dv' = 4\pi\left(\frac{\pi}{\omega}\right)^2 A g\int_0^{\tau/4} \sin\vartheta(v')\,g_1(v')\,dv'\,.$$

Für $n + m = 1$ ist $u_0 + u_0' = n$, so daß wegen

$$\frac{\vartheta_1\left(u_0' - v + \frac{\tau}{2}\right)}{\vartheta_0\left(u_0 + v - \frac{\tau}{2}\right)} = -\frac{\vartheta_0(u_0' + v)}{\vartheta_1(u_0 - v)}\,e^{-\pi i(u_0 + u_0')}$$

Gleichung (4.2) in

$$\omega(Z) = i\ln\frac{\vartheta_1(u_0 - v)\,\vartheta_1\left(u_0' - v + \frac{\tau}{2}\right)}{\vartheta_0(u_0' + v)\,\vartheta_0\left(u_0 + v - \frac{\tau}{2}\right)} = i\ln\frac{\vartheta_2(u_0 - v')\,\vartheta_2\left(u_0' - v' + \frac{\tau}{2}\right)}{\vartheta_3(u_0' + v')\,\vartheta_3\left(u_0 + v' - \frac{\tau}{2}\right)}$$

umgeformt werden kann. Mit (8.1) ergibt sich

$$e^{i\vartheta} = \frac{\vartheta_3(u_0' + v')\,\vartheta_3\left(u_0 + v' - \frac{\tau}{2}\right)}{\vartheta_2(u_0 - v')\,\vartheta_2\left(u_0' - v' + \frac{\tau}{2}\right)} = e^{\frac{4\pi i}{\tau}(u_0 + u_0')\left(\frac{\tau}{4} - v'\right)}\,\{1 + O(1)\}\,,$$

so daß $\sin\vartheta(v') = \sin\frac{\pi}{\tau}4(u_0+u_0')\left(\frac{\tau}{4}-v'\right)\{1+O(1)\}$ wird. Ferner ist

$$g_1(v') = e^{-\frac{2\pi i}{\tau}\{u_0^2+u_0'^2-3v_2^2-v_1'^2\}}\frac{\sin\frac{i\pi}{\tau}v'\cos\frac{i\pi}{\tau}v'}{\sin\frac{i\pi}{\tau}\left(\frac{\tau}{4}-v'\right)\sin\frac{i\pi}{\tau}\left(\frac{\tau}{4}+v'\right)}\{1+O(1)\},$$

da $v_1' = \frac{\tau}{4}$ wegen $s_0 = s_0'$ gilt. Daher folgt mit (8.3) $I \to 0$ für $t \to 0$.

b) Für $q \to +0$ erhält man mit Hilfe der Thetareihen Formeln, die die Strömung um ein einziges Hindernis charakterisieren. Aus (7.6) entsteht

$$e^{-\tau(e^{is})} = \left|\frac{\sin\frac{s+s_0}{2}}{\sin\frac{s-s_0}{2}}\right|, \qquad s_0 = n\pi.$$

Da $v_2 = v_1' = \frac{\tau}{4}$ für $s_0 = s_0'$ gilt, hat man für den Faktor in (8.4)

$$\lim_{q\to 0}\frac{\vartheta_2\left(u_0-\frac{\tau}{4}\right)\vartheta_2\left(u_0+\frac{\tau}{4}\right)\vartheta_3\left(u_0-\frac{\tau}{4}\right)\vartheta_3\left(u_0+\frac{\tau}{4}\right)}{\vartheta_2^3(0)\,\vartheta_2^3\left(\frac{\tau}{2}\right)} = \frac{1}{8},$$

während man für den Faktor $g_1(u)$ in (8.2)

$$\lim_{q\to+0}\frac{1}{q}g_1(u) = 4\sin s(\cos s - \cos s_0)$$

bekommt. Deshalb wird $\lim\limits_{q\to+0}\psi_1 = \infty$ und damit auch

$$\lim_{q\to+0}K(q, s_0) = \infty.$$

9.

Lemma: Das Gebiet D vom Flächeninhalt σ sei konform auf ein Rechteck, das Rechteck konform auf die obere F-Halbebene abgebildet; $\overline{F_1F_2}$ und $\overline{F_3F_4}$ seien zwei durchschnittsfremde Segmente der reellen Achse der F-Ebene, wobei die Punkte F_1, F_2, F_3, F_4 den Ecken $\omega+\omega'$, ω', 0, ω des Rechtecks entsprechen. Ist S die Länge der kürzesten Verbindung in D, die das Bild von $\overline{F_1F_2}$ mit dem Bild von $\overline{F_3F_4}$ verbindet, so gilt

$$\frac{S^2}{\sigma} \leqq \frac{1}{i}\frac{\omega'}{\omega}.$$

Bedeutet (F_1, F_3, F_4, F_2) das Doppelverhältnis der Punkte F_1, F_3, F_4, F_2, so wird

$$-(F_1, F_3, F_4, F_2) \geqq \eta\left[i\frac{S^2}{\sigma}\right]$$

mit einer monoton wachsenden, stetigen Funktion $\eta(\varepsilon)$, die für $\varepsilon = 0$ verschwindet. Ist $\overline{\overline{S}}$ die Länge der kürzesten Verbindung in D, die das Bild von $\overline{F_2F_3}$ mit dem Bild von $\overline{F_4F_1}$ verbindet, so hat man die Ungleichung

$$\frac{s^2}{\sigma} \leqq i\frac{\omega}{\omega'}.$$

Zum Beweis siehe [3], [4].

Satz: Für alle eventuellen Lösungen des Totwasserproblems zu vorgegebenen Winkeln n, m und Werten l, l', k mit $0 < n_0 < n$, $m < 1 - n_0 < 1$, $0 < l_0 < l$, $l' < l_1$ liegt der Parameter q in dem Intervall $0 < q_0 < q < q_1 < < 1$, wobei q_0, q_1 nur von n_0, l_0, l_1, k abhängt.

Beweis:

a) Im Schnittpunkt P der Parallelen durch P_1 zur y-Achse und durch P_2' zur x-Achse berühre eine Kugel vom Durchmesser $k + l + l'$ die z-Ebene. D sei die stereographische Projektion des Strömungsgebiets auf die Kugel vom Flächeninhalt $\sigma < \pi(k + l + l')^2$. Die Bilder der Segmente $\overline{F_1F_2}$ und $\overline{F_3F_4}$ (unteres bzw. oberes Hindernis) liegen dann auf der Kugel unterhalb des Äquators, und ihre kürzeste Verbindung wird auf dem Äquator erreicht. Wegen des Zusammenhanges der stereographischen mit den kartesischen Koordinaten folgt $S > \frac{1}{2} k$, so daß nach dem Lemma

$$q = e^{i\pi\frac{\omega'}{\omega}} < e^{-\frac{\pi}{4}\left(\frac{k}{k+l+l'}\right)^2} < e^{-\frac{\pi}{4}\left(\frac{k}{k+2l_1}\right)^2} = q_1 < 1$$

gilt.

b) Wird wie unter a) das Strömungsgebiet auf die Kugel projiziert, so muß jede im Strömungsgebiet verlaufende Verbindung eines Punktes des Bildes von $\overline{F_2F_3}$ mit einem Punkt des Bildes von $\overline{F_4F_1}$ den Äquator passieren. Da die beiden Hindernisse in der z-Ebene in Parallelstreifen der Breite $l' \cdot \sin m\pi$ bzw. $l \cdot \sin n\pi$ liegen, gilt

$$s > \frac{1}{2} \min \{l' \sin m\pi, l \sin n\pi\} > \frac{1}{2} l_0 \sin n_0\pi$$

und nach dem Lemma folgt

$$q = e^{i\pi\frac{\omega'}{\omega}} > e^{-\frac{4\pi(k+2l_1)^2}{l_0^2 \sin^2 n_0\pi}} = q_0 > 0 .$$

Satz: Für alle eventuellen Lösungen des Totwasserproblems, bei denen der horizontale Abstand h der beiden Ablösungspunkte P_1 und P_2' die Bedingung $|h| < H$ erfüllt, liegen die Nullstellen X_1, X_2 der Funktion $\vartheta(X)$ in Intervallen

$$q + \delta < X_2 < 1 - \delta , \qquad -1 + \delta < X_1 < -q - \delta , \qquad \delta > 0 ,$$

wobei δ nur von l_1, k, H abhängt.

Beweis:

a) Es sei $F_1 = -\infty$, $F_2 = d$, $F_3 = 1$, $F_4 = e$; den Segmenten $\overline{F_1F_2}$ bzw. $\overline{F_3F_4}$ entsprechen in der z-Ebene die untere freie Linie λ_2 bzw. das obere Hindernis. Dann wird

$$-(F_1, F_3, F_4, F_2) = \frac{1-d}{1-e} = \frac{e_3 - e_1}{\wp\left(\omega' - \frac{\omega}{i\pi} \ln X_2\right) - e_3} .$$

Eine Kugel vom Durchmesser $H + k + l + l' < H + k + 2l_1$ berühre die z-Ebene im Punkt P_2. Das obere Hindernis gehört zu einem Gebiet der Kugel, das unterhalb des Äquators zwischen den Bildern der Parallelen zur x-Achse durch P_1' bzw. P_2' liegt, während λ_2 zum Bild des Gebietes unterhalb der Parallelen durch P_2 gehört. Der Abstand dieser Gebiete ist mindestens so groß wie die Länge des Äquatorbogens zwischen den Schnittpunkten des Äquators mit den Bildern der Parallelen zur x-Achse durch P_1' und P_2'. Daraus folgt $S > \frac{1}{2} k$, und nach dem Lemma gilt für alle h mit $|h| < H$

$$\frac{e_3 - e_1}{\wp\left(\omega' - \frac{\omega}{i\pi} \ln X_2\right) - e_3} > \eta\left[i \frac{k^2}{4\pi(H + k + 2l_1)^2}\right].$$

Wegen $\lim\limits_{X_2 \to q} \wp\left(\omega' - \frac{\omega}{i\pi} \ln X_2\right) = \infty$ folgt daraus $X_2 > q + \delta$; $\delta > 0$.

b) Es sei $F_1 = d$, $F_2 = -1$, $F_3 = e$, $F_4 = \infty$; den Segmenten $\overline{F_1F_2}$ bzw. $\overline{F_3F_4}$ entsprechen in der z-Ebene das untere Hindernis bzw. die obere freie Linie λ_1'. Es ist

$$-(F_1, F_3, F_4, F_2) = -\frac{e+1}{d+1} = \frac{e_3 - \wp\left(\omega' - \frac{\omega}{i\pi} \ln X_2\right)}{e_3 - e_2} .$$

Berührt eine Kugel vom Durchmesser $H + k + l + l'$ im Punkt P_2 die z-Ebene, so gehören die Bilder der Segmente $\overline{F_1F_2}$ und $\overline{F_3F_4}$ wie unter a) zu Gebieten, die mindestens den Abstand $\frac{1}{2} k$ besitzen. Daher folgt

$$\frac{\wp\left(\omega' - \frac{\omega}{i\pi} \ln X_2\right) - e_3}{e_3 - e_2} > \eta\left[i \frac{k^2}{4\pi(H + k + 2l_1)^2}\right]$$

und wegen $\lim\limits_{X_2 \to 1} \wp\left(\omega' - \frac{\omega}{i\pi} \ln X_2\right) = e_3$ ist $X_2 < 1 - \delta$.

c) Es sei $F_1 = d$, $F_2 = -1$, $F_3 = b$, $F_4 = 1$ bzw. $F_2 = -1$, $F_2 = b$, $F_3 = 1$, $F_4 = e$; im ersten Fall ist

$$r_1 = -(F_1, F_3, F_4, F_2) = -\frac{d-1}{b-1}\frac{b+1}{d+1} = \frac{e_3 - e_1}{e_3 - e_2} \frac{e_2 - \wp\left(\omega' - \frac{\omega}{i\pi} \ln X_1\right)}{\wp\left(\omega' - \frac{\omega}{i\pi} \ln X_1\right) - e_1},$$

im zweiten Fall

$$r_2 = -(F_1, F_3, F_4, F_2) = -\frac{e+1}{e-1}\frac{b-1}{b+1} = \frac{\wp\left(\omega' - \frac{\omega}{i\pi}\ln X_1\right) - e_1}{e_2 - \wp\left(\omega' - \frac{\omega}{i\pi}\ln X_1\right)}.$$

Wird

$$A = \frac{\wp\left(\omega' - \frac{\omega}{i\pi}\ln X_2\right) - e_3}{e_3 - e_1}, \qquad B = \frac{\wp\left(\omega' - \frac{\omega}{i\pi}\ln X_1\right) - e_1}{e_1 - \wp\left(\omega' - \frac{\omega}{i\pi}\ln X_1\right)}$$

gesetzt, so ist für $A \leqq B$ nach a)

$$r_1 = \frac{e_3 - e_1}{e_3 - e_2} B \geqq \frac{\wp\left(\omega' - \frac{\omega}{i\pi}\ln X_2\right) - e_2}{e_3 - e_1} > \eta\left[i\frac{k^2}{4\pi(H + k + 2l_1)^2}\right]$$

und für $\frac{1}{A} \leqq \frac{1}{B}$ nach b)

$$r_2 = \frac{1}{B} \geqq \frac{e_3 - e_1}{\wp\left(\omega' - \frac{\omega}{i\pi}\ln X_2\right) - e_3} > \eta\left[i\frac{k^2}{4\pi(H + k + 2l_1)^2}\right].$$

Wegen $\lim\limits_{X_2 \to -1} \wp\left(\omega' - \frac{\omega}{i\pi}\ln X_1\right) = e_2$, $\lim\limits_{X_1 \to -q} \wp\left(\omega' - \frac{\omega}{i\pi}\ln X_1\right) = e_1$ folgt daher $-1 + \delta < X_1 < -q - \delta$.

Folgerung: Die zu den Nullstellen X_1, X_2 gehörenden Werte des Parameters s_0 liegen für alle α und q mit

$$0 \leqq \alpha \leqq (1 - 2n_0)\pi \quad (\text{bzw. } (2n_0 - 1)\pi \leqq \alpha \leqq 0), \qquad q_0 < q < q_1$$

in einem Intervall $0 < \varepsilon < s_0 < \pi - \varepsilon$, wobei ε nur von n_0, l_0, l_1, k, H abhängt.

Beweis: Es sei $\alpha_1 = \frac{\pi - \alpha}{4\pi}$ im Fall $0 \leqq \alpha \leqq (1 - 2n_0)\pi$; dann kann man wegen $u_0 + u_0' = \frac{\pi - \alpha}{2\pi} u_0 = \alpha_1 + u_1$, $u_0' = \alpha_1 - u_1$ mit $-\alpha_1 \leqq u_1 \leqq \alpha_1$, $\frac{1}{2} n_0 \leqq \alpha_1 \leqq \frac{1}{4}$ setzen. Aus $\omega(X_2) = 0$ erhält man mittels (4.2)

$$\frac{\vartheta_1(u_0 - v_2)\,\vartheta_0(u_0' - v_2)}{\vartheta_1(u_0 + v_2)\,\vartheta_0(u_0' + v_2)} = \frac{\vartheta_1(\alpha_1 - v_2 + u_1)\,\vartheta_0(\alpha_1 - v_2 - u_1)}{\vartheta_1(\alpha_1 + v_2 + u_1)\,\vartheta_0(\alpha_1 + v_2 - u_1)} = e^{i\pi(1-n)}.$$

Verwendet man die Beziehung

$$\vartheta_2\vartheta_3\vartheta_1(u + v)\,\vartheta_1(u - v) = \vartheta_1(u)\,\vartheta_0(u)\,\vartheta_2(v)\,\vartheta_3(v) + \vartheta_2(u)\,\vartheta_3(u)\,\vartheta_1(v)\,\vartheta_0(v),$$

so bekommt man zwischen u_1 und v_2 den Zusammenhang

$$-\frac{\vartheta_1(u_1)\,\vartheta_0(u_1)}{\vartheta_2(u_1)\,\vartheta_3(u_1)} = \frac{\vartheta_1(\alpha_1 + v_2)\,\vartheta_0(\alpha_1 + v_2)\,e^{i\pi(1-n)} - \vartheta_1(\alpha_1 - v_2)\,\vartheta_0(\alpha_1 - v_2)}{\vartheta_2(\alpha_1 + v_2)\,\vartheta_3(\alpha_1 + v_2)\,e^{i\pi(1-n)} - \vartheta_2(\alpha_1 - v_2)\,\vartheta_3(\alpha_1 - v_2)}$$

und eine analoge Beziehung zwischen u_2 und v_2. Für $X_2 \to 1 (v_2 \to 0)$ bzw. $X_1 \to -1$ $(v_1 \to \frac{1}{2})$ entnimmt man daraus $s_0 \to 0$ bzw. $s_0 \to \pi$. Da für die Nullstellen X_1, X_2 der eventuellen Lösungen $X_2 < 1 - \delta$, $X_1 > -1 + \delta$ gilt, existiert ein $\varepsilon > 0$, so daß die Parameterwerte s_0 in $\varepsilon < s_0 < \pi - \varepsilon$ liegen, und ε ist wegen der stetigen Abhängigkeit von α_1 und q in $\frac{1}{2} n_0 \leqq \alpha_1 \leqq \frac{1}{4}$, $q_0 < q < q_1$ eine Funktion der angegebenen Größen.

10.

Satz: Zwei geradlinige Hindernisse mit den Winkeln $m\pi$ bzw. $n\pi$ zur x-Achse, den Längen l' bzw. l und dem vertikalen Abstand k der beiden Ablösungspunkte P_1 und P_2' seien vorgegeben. Es gelte $n_0 < n$, $m < 1 - n_0$, $0 < n_0 < \frac{1}{2}$, $l_0 < l, l' < l_1$, $0 < k < \infty$. Dann existiert zu $H > 0$ eine Lösung des Totwasserproblems bezüglich dieser Vorgaben, wobei für den horizontalen Abstand h der Punkte P_1 und P_2' $|h| < H$ gilt.

Beweis: Es werden die Funktion $\mathcal{L}(q, s_0, l')$ und $\mathcal{K}(q, s_0, k)$ zu den Vorgaben $m_t = mt + \frac{1}{2}(1 - t)$, $n_t = nt + \frac{1}{2}(1 - t)$, $l_t' = l't + l(1 - t)$ betrachtet. Die Darstellungen durch Thetafunktionen zeigen, daß sie für $0 < q < 1$, $0 < s_0 < \pi - \alpha$ stetig von den Argumenten abhängen. Da $\omega(Z)$ stetig von den Randwerten abhängt, sind sie gleichgradig stetig in t. Wie im Punkt 9 gezeigt, gehören die eventuellen Lösungen zu Intervallen $\varepsilon < s_0 < \pi - \varepsilon$, $q_0 < q < q_1$, die nur von n_0, l_0, l_1, k, H abhängen, so daß für kein t eine Lösung auf dem Rand liegt. Für $t = 0$ hat Gleichung (6.5) genau eine Lösung $s_0 = s_0' = \pi/2$ (Punkt 7), während $0 < K(q, s_0) < \infty$ nach Punkt 8 für $1 > q > 0$ gilt. Der Totalindex der Lösungen der Gleichungen (6.5) und (6.6) ist daher für $t = 0$ von Null verschieden und damit auch für $t = 1$: es existiert eine Lösung des Totwasserproblems.

Literatur

1 BERG, P., The existence of subsonic Helmholtz flows of a compressible fluid, Comm. Pure Appl. Math. **15** (1962) 289–347.
2 DEMTSCHENKO, Problèmes mixtes harmoniques en hydrodynamique des fluides parfaits, Gauthier-Villars, Paris, 1933.
3 KRAVTSCHENKO, M. J., Sur le problème de représentation conforme de Helmholtz, theorie des sillages et des proues, Journal de mathematiques **22** (1943) 245 und **23** (1944) 1.
4 LERAY, J., Les problèmes de représentation conforme d'Helmholtz; théorie des sillages et des proues, Commentarii mathematici Helvetici **8** (1935) 149.
5 MAGNUS, W.; OBERHETTINGER, F., Formeln und Sätze für die speziellen Funktionen der mathematischen Physik, Springer-Verlag 1948.
6 SERRIN, J., Existence theorems for some hydrodynamical free boundary problems, J. Rational Mech. Anal. **1** (1952) 1–48.
7 TANNORY, I., MOLK, I., Eléments de la théorie de fonctions elliptiques, Gauthier-Villars, Paris 1896.
8 THIRY, R., Sur solutions multiples des problèmes d'hydrodynamique relatifs aux mouvements glissants, Annales sc. l'Ecole norm. sup. **38** (1921), 229.
9 VILLAT, M. H., Le problème de Dirichlet dans une aire annulaire, Rendiconti del Circolo Palermo **23** (1912) 134.
10 VILLAT, M. H., Sur le mouvement d'un solide donné dans un fluide limité par une paroi fixe, Journal de mathematiques **7** (1911) 353.

Eingereicht am 3. 8. 1979

Anschrift: Dr. HARALD HILBIG, Karl-Marx-Universität, Sektion Mathematik, 7010 Leipzig, Karl-Marx-Platz, DDR

ZAMM 61, 193–199 (1981)

N. PHAN-THIEN

On Stokes Flows in Channels and Pipes with Parallel Stationary Random Surface Roughness

The Stokes flow problems of an incompressible viscous fluid in corrugated channels and pipes are considered. The flow is parallel to the corrugations which are of small amplitude ε and are represented by stationary random functions. A pertubation solution is developed leading to an expression for the flow rate enhancement as a function of the statistics of the corrugations. Simplification of this expression are discussed.

Das Stokessche Strömungsproblem eines inkompressiblen, viskosen Mediums in geriffelten Kanälen und Rohren wird betrachtet. Die Strömung verläuft parallel zur Riffelung, die von der kleinen Amplitude ε ist und durch stationäre Zufallsfunktionen dargestellt wird. Eine Störungslösung wird entwickelt, die zu einem Ausdruck für das Anwachsen der Strömungsgeschwindigkeit als Funktion der Statistik der Riffelung führt. Vereinfachungen dieses Ausdrucks werden diskutiert.

Рассматривается задача Стокса течения несжимаемой вязкой жидкости в каналах и трубах с шероховатой поверхностью. Течение параллельно шероховатостью амплитуды ε, которая характеризуется стационарными случайными функциями. Получено возмущенное решение, которое ведет к выражению для повышения скорости течения как функция статистики шероховатости. Обсуждены упрощения этого выражения.

Introduction

The problem of STOKES flow through corrugated channels and pipes has been recently considered by CHOW and SODA [1], WANG [2–3] and PHAN-THIEN [4], for the case of corrugated channels, and PHAN-THIEN [5] for the case of corrugated pipes. Except in [4–5], all of these authors assumed a sinusoidal corrugation. In particular WANG has considered the flow transverse to the corrugations [2] and parallel to them [3]. Both studies show that a flow rate enhancement is possible and is of order $O(\varepsilon^2)$, where $\varepsilon \ll 1$ is the dimensionless amplitude of the corrugations. Moreover, the flow rate enhancement is a function of the frequency and the relative phase shift of the corrugations. Thus, by varying the relative phase shift, one can, to a limited extent, control the flow through corrugated plates. WANG's [2] results for the flows transverse to the corrugations have since been extended to channels [4] and pipes [5] with stationary random surface roughness.

In this communication, WANG's [3] results for flows parallel to the corrugations are generalized to channels and pipes with stationary random surface roughness. Any other deterministic surface roughness which satisfies DIRICHLET's condition can be regarded as a special case of this.

Channel Flow

We consider here the STOKES flow of an incompressible viscous fluid under a constant pressure gradient between two corrugated plates whose positions are described by

$$y = a\varepsilon m_1(z)\,, \qquad \text{and} \qquad y = a[1 + \varepsilon m_2(z)]\,. \tag{1}$$

Here a is a length scale, $\varepsilon \ll 1$ is a perturbation parameter, $m_k(z)$, $k = 1, 2$, are some random processes and x is the direction of flow (Figure 1a).

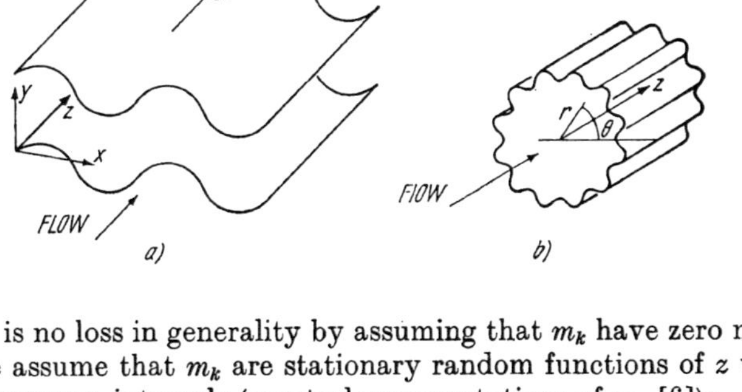

Fig. 1. Flow geometry (addendum in the proof-sheet: change x and z in Fig. 1a)

There is no loss in generality by assuming that m_k have zero means; thus a is the mean width of the channel. Further, we assume that m_k are stationary random functions of z whence they can be expressed by the following FOURIER-STIELTJES integrals (spectral representations of m_k [6]):

$$m_1(z) = \int_{-\infty}^{\infty} e^{i\lambda z}\, dZ_1(\lambda)\,; \qquad m_2(z) = \int_{-\infty}^{\infty} e^{i\lambda z}\, dZ_2(\lambda)\,, \tag{2}$$

where $i^2 = -1$ and $dZ_k(\lambda)$; $k = 1, 2$, are interval random functions [6] that satisfy

$$\langle dZ_k(\lambda)\rangle = 0\,; \qquad k = 1, 2\,; \qquad l = 1, 2\,, \tag{3}$$

$$\langle dZ_k(\lambda_i)\, \overline{dZ_l(\lambda_j)}\rangle = \delta_{ij}\Omega_{kl}(\lambda_i)\, d\lambda_i\,, \qquad \text{(no sum)} \tag{4}$$

where the overbar denotes a complex conjugate, $\langle\cdot\rangle$, an ensemble average, δ_{ij} is the KRONECKER delta and $\Omega_{kl}(\lambda)$ are the spectral densities defined by

$$R_{kl}(s) = \langle m_k(x+s)\, m_l(x)\rangle = \int_{-\infty}^{\infty} e^{i\lambda z}\, \Omega_{kl}(\lambda)\, d\lambda\,, \tag{5}$$

or equivalently,

$$\Omega_{kl}(\lambda) = \frac{1}{2\pi} \int_{-\infty}^{\infty} e^{-i\lambda s}\, R_{kl}(s)\, ds\,. \tag{6}$$

Note that we have tacitly assumed that $|R_{kl}(s)|$ tend to zero fast enough as $s \to \infty$ so that they all possess FOURIER transforms at least in the generalized sense.

The relevant STOKES equation for this flow is

$$\frac{\partial^2}{\partial y^2} u + \frac{\partial^2}{\partial z^2} u = -\frac{P}{\eta}, \tag{7}$$

where $u = u(y, z)$ is the axial velocity, P is the pressure drop per unit length and η is the fluid viscosity. Note that continuity requires u to be invariant in x.

The imposed boundary conditions are:

$$\text{at} \quad y = a\varepsilon m_1(z)\,, \qquad u = 0; \tag{8}$$

$$\text{at} \quad y = a[1 + \varepsilon m_2(z)]\,, \qquad u = U\,. \tag{9}$$

We seek a perturbation solution of the form

$$u = u_0 + \varepsilon u_1 + \varepsilon^2 u_2 + O(\varepsilon^3)\,, \tag{10}$$

from which the boundary condition (8) becomes

$$0 = u_0(0) + \varepsilon \left\{u_1(0) + \varepsilon a m_1 \frac{\partial u_0}{\partial y}(0)\right\} + \varepsilon^2 \left\{u_2(0) + \varepsilon a m_1 \frac{\partial u_1}{\partial y}(0) + \frac{1}{2}\varepsilon^2 a^2 m_1^2 \frac{\partial^2 u_0}{\partial y^2}(0)\right\} + O(\varepsilon^3) \tag{11}$$

and a similar expression for (9). In (11), the argument "0" denotes an evaluation at $y = 0$.

The ε^0-solution is the usual COUETTE-POISEUILLE profile:

$$u_0 = \frac{P}{2\eta} y(a-y) + \frac{U}{a} y\,. \tag{12}$$

The ε^1-solution is governed by

$$\frac{\partial^2}{\partial y^2} u_1 + \frac{\partial^2}{\partial z^2} u_1 = 0\,, \tag{13}$$

where the boundary conditions for u_1 are

$$u_1(0) = -a m_1 \frac{\partial u_0}{\partial y}(0) = -\left(\frac{Pa^2}{2\eta} + U\right) m_1\,, \tag{14}$$

and

$$u_1(a) = -a m_2 \frac{\partial u_0}{\partial y}(a) = \left(\frac{Pa^2}{2\eta} - U\right) m_2\,. \tag{15}$$

In view of (14—15), the spectral representation of u_1 is

$$u_1 = \int_{-\infty}^{\infty} \varphi_k(y, \lambda)\, e^{i\lambda z}\, dZ_k(\lambda)\,, \tag{16}$$

where the usual summation convention has been used and φ_k, $k = 1, 2$, satisfy

$$\frac{\partial^2}{\partial y^2}\varphi_k - \lambda^2 \varphi_k = 0\,, \tag{17}$$

where

$$\varphi_1(0, \lambda) = -\left(\frac{Pa^2}{2\eta} + U\right), \qquad \varphi_1(a, \lambda) = 0; \tag{18}$$

$$\varphi_2(0, \lambda) = 0\,, \qquad \varphi_2(a, \lambda) = \frac{Pa^2}{2\eta} - U\,. \tag{19}$$

The solutions of (17—19) are easily determined and we obtain

$$u_1 = \left(\frac{Pa^2}{2\eta} + U\right) \int_{-\infty}^{\infty} \frac{\sinh \lambda(y-a)}{\sinh \lambda a}\, e^{i\lambda z}\, dZ_1(\lambda) + \left(\frac{Pa^2}{2\eta} - U\right) \int_{-\infty}^{\infty} \frac{\sinh \lambda y}{\sinh \lambda a}\, e^{i\lambda z}\, dZ_2(\lambda)\,. \tag{20}$$

It should be noted that $\langle u_1 \rangle = 0$ and any quantity depending linearly on u will have its mean to be a function of y plus a term of the order $O(\varepsilon^2)$.

Next, the ε^2-solution is governed by

$$\frac{\partial^2}{\partial y^2} u_2 + \frac{\partial^2}{\partial z^2} u_2 = 0 \tag{21}$$

where

$$\begin{aligned} u_2(0) &= -am_1 \frac{\partial u_1}{\partial y}(0) - \frac{1}{2} a^2 m_1^2 \frac{\partial^2 u_0}{\partial y^2}(0) \\ &= -\left(\frac{Pa^2}{2\eta} + U\right) \int_{-\infty}^{\infty}\int_{-\infty}^{\infty} a\lambda \coth(a\lambda)\, e^{i(\lambda-\lambda')z} dZ_1(\lambda)\, \overline{dZ_1}(\lambda') - \\ &\quad -\left(\frac{Pa^2}{2\eta} - U\right) \int_{-\infty}^{\infty}\int_{-\infty}^{\infty} \frac{a\lambda}{\sinh a\lambda} e^{i(\lambda-\lambda')z}\, dZ_2(\lambda)\, \overline{dZ_1}(\lambda') + \frac{Pa^2}{2\eta} \int_{-\infty}^{\infty}\int_{-\infty}^{\infty} e^{i(\lambda-\lambda')z}\, dZ_1(\lambda)\, \overline{dZ_1}(\lambda') \end{aligned} \tag{22}$$

and

$$\begin{aligned} u_2(a) &= -am_2 \frac{\partial u_1}{\partial y}(a) - \frac{1}{2} m_2^2 a^2 \frac{\partial^2 u_0}{\partial y^2}(a) = -\left(\frac{Pa^2}{2\eta} + U\right) \int_{-\infty}^{\infty}\int_{-\infty}^{\infty} \frac{a\lambda}{\sinh \lambda a} e^{i(\lambda-\lambda')z}\, dZ_1(\lambda)\, \overline{dZ_1}(\lambda') - \\ &\quad -\left(\frac{Pa^2}{2\eta} - U\right) \int_{-\infty}^{\infty}\int_{-\infty}^{\infty} a\lambda \coth(a\lambda)\, e^{i(\lambda-\lambda')}\, dZ_2(\lambda)\, \overline{dZ_2}(\lambda') + \frac{Pa^2}{2\eta} \int_{-\infty}^{\infty}\int_{-\infty}^{\infty} e^{i(\lambda-\lambda')z}\, dZ_2(\lambda)\, \overline{dZ_2}(\lambda'). \end{aligned} \tag{23}$$

The boundary conditions suggest a spectral representation for u_2 of the form

$$u_2 = \int_{-\infty}^{\infty}\int_{-\infty}^{\infty} \zeta_{kl}(y, \lambda, \lambda')\, e^{i(\lambda-\lambda')z}\, dZ_k(\lambda)\, \overline{dZ_l}(\lambda') \tag{24}$$

where

$$\frac{\partial^2}{\partial y^2} \zeta_{kl} - (\lambda - \lambda')^2 \zeta_{kl} = 0\,, \tag{25}$$

and

$$\zeta_{11}(0, \lambda, \lambda') = \frac{Pa^2}{2\eta} [1 - a\lambda \coth(a\lambda)] - Ua\lambda \coth(a\lambda)\,, \tag{26}$$

$$\zeta_{11}(a, \lambda, \lambda') = \zeta_{21}(a, \lambda, \lambda') - \zeta_{12}(0, \lambda, \lambda') = \zeta_{22}(0, \lambda, \lambda') = 0\,, \tag{27}$$

$$\zeta_{21}(0, \lambda, \lambda') = -\left(\frac{Pa^2}{2\eta} - U\right) \frac{a\lambda}{\sinh a\lambda}, \tag{28}$$

$$\zeta_{12}(a, \lambda, \lambda') = -\left(\frac{Pa^2}{2\eta} - U\right) \frac{a\lambda}{\sinh a\lambda}, \tag{29}$$

$$\zeta_{22}(a, \lambda, \lambda') = \frac{Pa^2}{2\eta} [1 - \coth(a\lambda)] + Ua\lambda \coth(a\lambda)\,. \tag{30}$$

The solutions to (25—30) are easily found and we finally obtain

$$\begin{aligned} u_2 &= \frac{Pa^2}{2\eta} \int_{-\infty}^{\infty}\int_{-\infty}^{\infty} \Bigg\{ (1 - a\lambda \coth \lambda a) \frac{\sinh(\lambda - \lambda')(h_0 - y)}{\sinh(\lambda - \lambda') h_0} dZ_1(\lambda)\, \overline{dZ_1}(\lambda') - \\ &\quad - \frac{\lambda a}{\sinh \lambda a} \left(\cosh(\lambda - \lambda') y - \frac{\sinh(\lambda - \lambda') y}{\sinh(\lambda - \lambda') a} [\cosh(\lambda - \lambda') a - 1] \right) dZ_1(\lambda)\, \overline{dZ_2}(\lambda') + \\ &\quad + (1 - a\lambda \coth \lambda a) \frac{\sinh(\lambda - \lambda') y}{\sinh(\lambda - \lambda') a} dZ_2(\lambda)\, \overline{dZ_2}(\lambda)' \Bigg\} e^{i(\lambda-\lambda')z} - \\ &\quad - U \int_{-\infty}^{\infty}\int_{-\infty}^{\infty} \Bigg\{ a\lambda \coth \lambda a \frac{\sinh(\lambda - \lambda')(a - \lambda)}{\sinh(\lambda - \lambda') a} dZ_1(\lambda)\, \overline{dZ_1}(\lambda') - \\ &\quad - \frac{\lambda a}{\sinh \lambda a} \left(\cosh(\lambda - \lambda') y - \frac{\sinh(\lambda - \lambda') y}{\sinh(\lambda - \lambda') a} [\cosh(\lambda - \lambda') a + 1] \right) dZ_1(\lambda)\, \overline{dZ_2}(\lambda') - \\ &\quad - a\lambda \coth a\lambda \frac{\sinh(\lambda - \lambda') y}{\sinh(\lambda - \lambda') a} dZ_2(y)\, \overline{dZ_2}(\lambda') \Bigg\} e^{i(\lambda-\lambda')z}\,. \end{aligned} \tag{31}$$

The mean flow rate per unit width of the channel can be obtained as

$$\langle Q\rangle = \left\langle \int_{a\varepsilon m_1}^{a(1+\varepsilon m_2)} (u_0 + \varepsilon u_1 + \varepsilon^2 u_2)\, dy \right\rangle \tag{32}$$

The calculations are tedious and the final result is recorded here

$$\langle Q\rangle = \frac{Pa^2}{12\eta}(1 + \varepsilon^2 B_p) + \frac{1}{2} Ua(1 + \varepsilon^2 B_c), \tag{33}$$

where

$$B_p = \int_{-\infty}^{\infty} \left\{6 - 3a\lambda \coth(a\lambda)\right] [\Omega_{11}(\lambda) + \Omega_{22}(\lambda)] - \frac{a\lambda\, \Omega_{12}(\lambda)}{\sinh a\lambda}\right\} d\lambda \tag{34}$$

and

$$B_c = \int_{-\infty}^{\infty} (1 - a\lambda \coth a\lambda)\, [\Omega_{11}(\lambda) - \Omega_{22}(\lambda)]\, d\lambda . \tag{35}$$

It is interesting to note that, if the statistics of m_1 and m_2 are identical, then there is no flow enhancement for the COUETTE flow ($B_c = 0$).

Pipe Flow

Next, the pipe flow problem of an incompressible viscous fluid is considered (Figure 1b). The radius of the pipe is given by

$$R = a[1 + \varepsilon n(\theta)], \qquad \varepsilon \ll 1, \tag{36}$$

where n is an arbitrary random function of θ which has zero mean. Periodicity of $n(\theta)$ requires that $n(\theta + 2\pi k) = n(\theta)$, where k is an integer. This consideration leads to, if we assume that $n(\theta)$ is stationary,

$$n(\theta) = \sum_{k=-\infty}^{\infty} e^{ik\theta} Z(k), \qquad k = \text{integer}, \tag{37}$$

where $Z(k)$ is a random function of k which satisfies

$$\langle Z(k)\rangle = 0 \tag{38}$$

and

$$\langle Z(k)\, \bar{Z}(l)\rangle = \delta_{kl}\Omega(k), \quad \text{(no sum)}. \tag{39}$$

Here, Ω is related to the correlation function of $n(\theta)$ through

$$R(\psi) = \langle n(\theta + \psi)\, n(\theta)\rangle = \sum_{k=-\infty}^{\infty} e^{ik\psi}\, \Omega(k), \tag{40}$$

or, alternatively,

$$\Omega(k) = \frac{1}{2\pi} \int_0^{2\pi} R(\psi)\, e^{-ik\psi}\, d\psi . \tag{41}$$

Henceforth k and l denote integers and, without any loss of generality, we restrict θ to the range $[0, 2\pi]$.

The STOKES equation for this flow field is

$$\frac{1}{r}\frac{\partial}{\partial r} r \frac{\partial u}{\partial r} + \frac{1}{r^2}\frac{\partial^2 u}{\partial \theta^2} = -\frac{P}{\eta}, \tag{42}$$

where $u(r, \theta)$ is the axial velocity, P, the pressure drop per unit length, η, the fluid viscosity and (r, θ, z) are the cylindrical co-ordinates. Note that continuity requires u to be invariant in z.

The relevant boundary conditions for this flow are

$$u[a(1 + \varepsilon n)] = 0 \tag{43}$$

and that u must be bounded everywhere.

Again, we seek for a perturbation solution of the form

$$u = u_0 + \varepsilon u_1 + \varepsilon^2 u_2 + O(\varepsilon^3) \tag{44}$$

from which the boundary condition (43) becomes

$$u_0(a) + \varepsilon\left\{u_1(a) + an \frac{\partial u_0}{\partial r}(a)\right\} + \varepsilon^2\left\{u_2(a) + an \frac{\partial u_1}{\partial r}(a) + \frac{1}{2} a^2 n^2 \frac{\partial^2 u_0}{\partial r^2}(a)\right\} + O(\varepsilon^3) = 0 . \tag{45}$$

The ε^0-solution is the usual POISEUILLE profile:

$$u_0 = \frac{P}{4\eta}(a^2 - r^2) . \tag{46}$$

The ε^1-solution is governed by

$$\frac{1}{r}\frac{\partial}{\partial r}r\frac{\partial u_1}{\partial r}+\frac{1}{r^2}\frac{\partial^2 u_1}{\partial\theta^2}=0\,, \tag{47}$$

where

$$u_1(a)=-an\frac{\partial u_0}{\partial r}(a)=\frac{Pa^2}{2\eta}n\,. \tag{48}$$

This suggests a spectral representation for u_1 of the form

$$u_1=\frac{Pa^2}{2\eta}\sum_{k=-\infty}^{\infty}\varphi(r,k)\,e^{ik\theta}\,Z(k) \tag{49}$$

which requires that

$$\frac{1}{r}\frac{\partial}{\partial r}r\frac{\partial\varphi}{\partial r}-\frac{k^2}{r^2}\varphi=0\,, \tag{50}$$

where

$$\varphi(a,k)=1\,,\qquad |\varphi(r,k)|<\infty\,. \tag{51}$$

The solution to (50) is $(r/a)^{|k|}$ which yields

$$u_1=\frac{Pa^2}{2\eta}\sum_{k=-\infty}^{\infty}\left(\frac{r}{a}\right)^{|k|}e^{ika}\,Z(k)\,. \tag{52}$$

Next, the ε^2-solution is governed by

$$\frac{1}{r}\frac{\partial}{\partial r}r\frac{\partial u_2}{\partial r}+\frac{1}{r^2}\frac{\partial^2 u_2}{\partial\theta^2}=0 \tag{53}$$

where

$$u_2(a)=-an\frac{\partial u_1}{\partial r}(a)-\frac{1}{2}a^2n^2\frac{\partial^2 u_0}{\partial r^2}(a)=\frac{Pa^2}{4\eta}\sum_{k,l}(1-2|k|)\,e^{i(k-l)\theta}\,Z(k)\,\overline{Z}(l)\,. \tag{54}$$

This suggests a spectral representation for u_2 of the form

$$u_2=\frac{Pa^2}{4\eta}\sum_{k,l}\zeta(r,k,l)\,e^{i(k-l)\theta}\,Z(k)\,\overline{Z}(l)\,, \tag{55}$$

which requires that

$$\frac{1}{r}\frac{\partial}{\partial r}r\frac{\partial\zeta}{\partial r}-\frac{(k\quad l)^2}{r^2}\zeta=0\,, \tag{56}$$

where

$$\zeta(a,k,l)=1-2|k|\,,\qquad |\zeta(r,k,l)|<\infty\,. \tag{57}$$

Equations (56—57) can be easily integrated and we finally obtain

$$u_2=\frac{Pa^2}{4\eta}\sum_{k,l}(1-2|k|)\left(\frac{r}{a}\right)^{|k-l|}e^{i(k-l)\theta}\,Z(k)\,\overline{Z}(l)\,. \tag{58}$$

Finally the mean flow rate is given by

$$\langle Q\rangle=\langle 2\pi\int_0^{a(1+\varepsilon n)}[u_0+\varepsilon u_1+\varepsilon^2u_2+O(\varepsilon^3)]\,r\,dr\rangle=\frac{\pi Pa^4}{8\eta}(1+\varepsilon^2B_{pp})\,, \tag{59}$$

where

$$B_{pp}=2\sum_{k=-\infty}^{\infty}(3-2|k|)\,\Omega(k)\,. \tag{60}$$

Some Results

For sinusoidal corrugations we have $m_1=\sin\omega x$, $m_2=\sin(\omega x+\beta)$ and $n=\sin\omega\theta$. It is easy to show that their spectral densities are

$$\Omega_{11}(\lambda)=\Omega_{22}(\lambda)=\tfrac{1}{4}[\delta(\lambda-\omega)+\delta(\lambda+\omega)]\,,$$

$$\Omega_{12}(\lambda)=\frac{1}{4}\cos\beta[\delta(\lambda-\omega)+\sigma(\omega+\lambda)]+\frac{1}{4i}\sin\theta\,[\delta(\omega-\lambda)-\delta(\omega+\lambda)]$$

and

$$\Omega(k)=\tfrac{1}{2}\delta_{\omega|k|}\ (\omega\text{ is an integer in this case})\,.$$

Thus we have, by a direct substitution into (34), (35) and (60),

$$B_p = 6\left(1 - \lambda a \frac{\cosh \lambda a + \cos \beta}{\sinh \lambda a}\right), \tag{61}$$

as WANG [3] has shown by a different method and,

$$B_c = 0\,, \qquad B_{pp} = 3 - 2|\omega|\,, \qquad \omega\colon \text{integer}\,. \tag{62), (63}$$

We record here two approximations to (34—35):

Firstly, if the corrugations vary slowly then their spectral densities are mainly concentrated at a finite neighbourhood of $\lambda \sim 0$ (i.e. m_k are composed of low frequency signals).

Since

$$\Delta_1 \equiv 6 - 3a\lambda \coth a\lambda \sim 3 - (a\lambda)^2 + O(a\lambda)^4\,,$$

$$\Delta_2 \equiv a\lambda/\sinh a\lambda \sim 1 - \tfrac{1}{6}(a\lambda)^2 + O(a\lambda)^4\,,$$

and

$$\Delta_3 \equiv 1 - a\lambda \coth a\lambda \sim -\tfrac{1}{3}(a\lambda)^2 + O(a\lambda)^4\,,$$

we have

$$B_p = \int_{-\infty}^{\infty} \left\{[3 - (a\lambda)^2](\Omega_{11} + \Omega_{22}) - \left[1 - \frac{(a\lambda)^2}{6}\right]\Omega_{12}\right\} d\lambda + \dots$$

$$= 3(\langle m_1^2\rangle + \langle m_2^2\rangle) - a^2(\langle m_1'^2\rangle + \langle m_2'^2\rangle) - \langle mn\rangle + \frac{a^2}{6}\langle m'n'\rangle + E\,, \tag{64}$$

and

$$B_c = \int_{-\infty}^{\infty} -\frac{1}{3}(a\lambda)^2(\Omega_{11} - \Omega_{22})\, d\lambda + E = \frac{a^2}{3}(\langle m_2'^2\rangle - \langle m_1'^2\rangle) + E\,, \tag{65}$$

where E is an error term of the order $O(|\int a^4\lambda^4\, \Omega(\lambda)\, d\lambda|)$.

Lastly, if there is a dominant peak situated at ω in the spectra $\Omega_{kl}(\lambda)$ [the "natural" frequency of the corrugations] we may expand Δ_k, $k = 1, 2, 3$ in a TAYLOR's series about $\lambda = \omega$:

$$\Delta_k(a\lambda) = \Delta_k^{(0)} + a(\lambda - \omega)\,\Delta_k^{(0)'} + \tfrac{1}{2}a^2(\lambda - \omega)^2\,\Delta_k^{(0)''} + \dots,$$

where we have used the shorthand notation

$$\Delta_k^{(0)} = \Delta_k(a\omega)\,, \qquad \Delta_k^{(0)'} = \left.\frac{d\Delta_k(x)}{dx}\right|_{x=a\omega}, \quad \text{etc.}$$

Then, in this case we have

$$B_p = (\Delta_1^{(0)} + \tfrac{1}{2}a^2\omega^2\Delta_1^{(0)''})\langle m_1^2 + m_2^2\rangle - (\Delta_2^{(0)} + \tfrac{1}{2}a^2\omega^2\Delta_2^{(0)''})\langle m_1 m_2\rangle + \tfrac{1}{2}a^2\langle m_1'^2 + m_2'^2\rangle\,\Delta_1^{(0)''} - \tfrac{1}{2}a^2\Delta_2^{(0)''}\langle m_1' m_2'\rangle + R\,, \tag{66}$$

and

$$B_c = (\Delta_3^{(0)} + \tfrac{1}{2}a^2\omega^2\Delta_3^{(0)''})\langle m_1^2 - m_2^2\rangle + \tfrac{1}{2}a^2\Delta_3^{(0)''}\langle m_1'^2 - m_2'^2\rangle + R \tag{67}$$

where

$$R = O\left(\left|\int_{-\infty}^{\infty} a^4(\lambda - \omega)^4\,\Omega_{ij}(\lambda)\, d\lambda\right|\right).$$

To sum up we have developed a perturbation solution to the STOKES flow of an incompressible viscous fluid through a channel or a pipe with axial surface roughness. The corrugations on the walls are assumed to be of small amplitude and can be represented by stationary functions (for channels) and sequences (for pipes). This study points to the following conclusions:

The flow rate enhancement is an order $O(\varepsilon^2)$ effect and depends on the statistics of the corrugations in the manner indicated by (34—35) *and* (60).

For slowly varying corrugations, the flow rate enhancement depends only on the mean square value of m_k and n.

Finally if $m = \sum_{k=1}^{\infty} a_k \sin k\omega x$ *then its spectral density is*

$$\Omega(\lambda) = \tfrac{1}{4}\sum_{k=1}^{\infty} a_k^2[\delta(k\omega - \lambda) + \delta(k\omega + \lambda)]\,,$$

thus any deterministic corrugation shape which is Fourier series expressible (i.e. satisfies Dirichlet conditions) is a special of this.

Acknowledgement

It is a pleasure to thank the Australian Research Grants Committee for their support.

References

1 CHOW, J. C. F. and SODA, K., Laminar Flow and Blood Oxygeneration in Channels with Boundary Irregularities, J. Appl. Mech. ASME, **40**, 1973, pp. 843–850.
2 WANG, C.-Y., On STOKES Flow Between Corrugated Plates, J. Appl. Mech. ASME, **46**, 1979, pp. 462–464.
3 WANG, C.-Y., Parallel Flow Between Corrugated Plates, ASCE J. Eng. Mech., **102**, 1976, pp. 1088–1090.
4 PHAN-THIEN, N., On STOKES Flow Between Parallel Plates with Stationary Random Roughness, ZAMM, **60**, 675–679 (1980).
5 PHAN-THIEN, N., On STOKES Flow of a NEWTONian Fluid through a Pipe with Stationary Random Surface Roughness, Phys. Fluids, submitted.
6 YAGLOM, A. M., An Introduction to the Theory of Stationary Random Functions, translated by R. A. SILVERMAN, Prentice-Hall, 1965.

Eingereicht am 14. Mai 1980

Anschrift: Dr. N. PHAN-THIEN, Department of Mechanical Engineering, The University of Sydney, New South Wales, 2006, Australia

KLEINE MITTEILUNGEN

ZAMM 61, 201 (1981)

N. ANDERSON / A. M. ARTHURS

Variational Solution of a Nonlinear Boundary Value Problem in the Theory of Power Law Fluids

The nonlinear boundary value problem described by the BLASIUS-like equation

$$\frac{d^3w}{d\eta^3} + w(\eta)\left(\frac{d^2w}{d\eta^2}\right)^{2-n} = 0\,, \qquad n > 0\,, \tag{1}$$

with

$$w = \frac{dw}{d\eta} = 0 \quad \text{at} \quad \eta = 0\,, \qquad \frac{dw}{d\eta} \to 1 \quad \text{as} \quad \eta \to \infty\,, \tag{2}$$

arises in the study of the uniform flow of Newtonian ($n = 1$) and non-Newtonian ($n \neq 1$) fluids past a semi-infinite flat plate at zero incidence [cf. 1, 2]. With the introduction of CROCCO-like variables

$$x = w'(\eta)\,, \qquad V = w''(\eta)\,, \tag{3}$$

and the function φ defined by

$$\varphi = V^n\,, \tag{4}$$

the system in (1) and (2) becomes

$$-\varphi''(x) = nx\varphi^{-1/n}\,, \qquad 0 < x < 1\,, \tag{5}$$

with

$$\varphi'(0) = 0\,, \qquad \varphi(1) = 0\,. \tag{6}$$

We wish to consider the boundary value problem in this form as it is well suited to variational work. Of particular interest is $\varphi(0)$ which is closely related to the drag experienced by the fluid at the plate.

To give a variational formulation of (5) and (6) we first set

$$f(\varphi) = nx\varphi^{-1/n}\,, \tag{7}$$

and define

$$F(\varphi) = \int^{\varphi} f(v)\,dv = \frac{-n^2x}{1-n}\,\varphi^{1-(1/n)} \qquad (n \neq 1)\,,$$
$$= x \log \varphi \qquad (n = 1)\,. \tag{8}$$

We also define the integrals

$$J(\varphi_1) = \int_0^1 \{\tfrac{1}{2}\,\varphi_1'^2 - F(\varphi_1)\}\,dx\,, \qquad \varphi_1(1) = 0\,, \tag{9}$$

$$G(\psi_2) = \int_0^1 \{-\psi_2''f^{-1}(-\psi_2'') - \tfrac{1}{2}\,\psi_2'^2 - F[f^{-1}(-\psi_2'')]\}\,dx\,,$$
$$\psi_2'(0) = 0\,. \tag{10}$$

Then, since $F(\varphi)$ is concave or equivalently $f(\varphi)$ is antimonotone, it follows from the theory of complementary variational principles [cf. 3] that the solution φ of (5) and (6) minimizes $J(\varphi_1)$ and maximizes $G(\psi_2)$, viz.

$$G(\psi_2) \leqq G(\varphi) = J(\varphi) \leqq J(\varphi_1)\,, \tag{11}$$

equality arising when $\varphi_1 = \varphi$ and $\psi_2 = \varphi$. In addition a bound on the error in the variational function φ_1 can be derived from the complementary principles for J and G [cf. 3]. For our class of problems this is given by

$$\|\varphi_1 - \varphi\|_{L^2} \leqq E(\varphi_1)\,, \tag{12}$$

where $\|\psi\|_{L^2}$ is the L^2 norm $\left\{\int_0^1 \psi^2\,dx\right\}^{1/2}$, and

$$E(\varphi_1) = \left[\frac{8}{\pi^2}\{J(\varphi_1) - G(\psi_2)\}\right]^{1/2}, \tag{13}$$

this bound (12) holding for trials φ_1 which satisfy the exact boundary conditions in (6).

The Newtonian case $n = 1$ has already been solved variationally by the authors [4]. Here we examine the non-Newtonian cases $0 < n < 1$ (pseudoplastic fluids) and $n > 1$ (dilatant fluids). For these (9) and (10) become

$$J(\varphi_1) = \int_0^1 \left\{\frac{1}{2}\,\varphi_1'^2 + \frac{n^2x}{1-n}\,\varphi_1^{1-(1/n)}\right\} dx\,, \qquad \varphi_1(1) = 0\,, \tag{14}$$

$$G(\psi_2) = \int_0^1 \left\{\frac{(nx)^n}{1-n}\,(-\psi_2'')^{1-n} - \frac{1}{2}\,\psi_2'^2\right\} dx\,, \qquad \psi_2'(0) = 0\,. \tag{15}$$

Near $x = 0$ the exact solution φ of (5) and (6) has a power series expansion

$$\varphi = a_0 - \frac{na_0^{-1/n}}{3!}\,x^3 - \frac{5na_0^{-1-(2/n)}}{6!}\,x^6 + \dots\,, \tag{16}$$

and this suggests that we take trial functions of the form

$$\varphi_1 = \alpha_1(1 - x^3)^{\beta_1}\,, \tag{17}$$

$$\psi_2 = \alpha_2(1 - x^3)^{\beta_2}\,, \tag{18}$$

which satisfy the exact boundary conditions in (6). The parameters α and β were determined by optimizing the integrals J and G in (14) and (15), and the results for various values of n are given in Table 1. From these we see that the variational solution (17) provides quite an accurate representation of the function φ.

Table 1. Variational parameters

n	α_1	β_1	J	α_2	β_2	G	E
2.0	0.54	0.92	−0.8273	0.53	0.91	−0.8315	0.058
1.5	0.50	0.89	−1.221	0.49	0.88	−1.226	0.064
1.2	0.48	0.86	−2.642	0.48	0.85	−2.647	0.064
1.1	0.48	0.84	−5.109	0.48	0.84	−5.114	0.064
1.05	0.48	0.84	−10.09	0.48	0.83	−10.10	0.090
1.01	0.48	0.83	−50.07	0.48	0.82	−50.08	0.090
0.99	0.48	0.82	49.93	0.48	0.82	49.93	0.000
0.95	0.48	0.82	9.952	0.48	0.81	9.947	0.064
0.90	0.49	0.81	4.977	0.48	0.80	4.972	0.064
0.80	0.49	0.78	2.540	0.49	0.78	2.534	0.070
0.70	0.50	0.75	1.796	0.50	0.74	1.789	0.075
0.60	0.52	0.71	1.525	0.52	0.70	1.518	0.075
0.50	0.55	0.65	1.580	0.55	0.65	1.573	0.075
0.40	0.56	0.61	1.827	0.56	0.61	1.820	0.075

References

1 NACHMAN, A., and TALIAFERRO, S., Mass transfer into boundary layers for power law fluids, Proc. Roy. Soc. Lond. A 365, 313–326 (1979).
2 NACHMAN, A., and CALLEGARI, A., A nonlinear singular boundary value problem in the theory of pseudoplastic fluids, SIAM J. Appl. Math. 38, 275–281 (1980).
3 ARTHURS, A. M., Complementary variational principles, Clarendon Press, Oxford, 2nd ed., 1980.
4 ANDERSON, N., and ARTHURS, A. M., Extremum principles and error bounds for a nonlinear singular boundary value problem in the theory of viscous fluids, Nuovo Cimento Lett., 3, 562–566 (1972).

Eingereicht am 17. 7. 1980

Anschrift: Prof. A. M. ARTHURS, Department of Mathematics, University of York, Heslington, York, Y01 5DD, Great Britain

ZAMM 61, 202 – 203 (1981)

D. Karamanlidis

Über die numerische Berechnung ebener Rahmentragwerke im unter- und überkritischen Bereich nach der Finite-Element-Methode[1])

1. Einleitung

Die Berechnung des Tragverhaltens ebener Rahmentragwerke unter Berücksichtigung endlicher Deformationen ist bekanntlich nur in Sonderfällen geschlossen möglich, so daß in praxi meistens auf numerische Verfahren zurückgegriffen werden muß.

Die Finite-Element-Methode (FEM), die auch im Rahmen dieser Arbeit verwendet worden ist, hat sich seit den frühen 60er Jahren zu einem universell einsetzbaren Hilfsmittel zur Lösung der verschiedenartigen, nichtlinearen Randwertaufgaben der Ingenieurpraxis entwickelt. Grundsätzlich lassen sich drei Gesichtspunkte der Elementverfahren unterscheiden: a) die Schaffung einer geeigneten Diskretisierungsgrundlage für die einzelnen Elemente, b) der Zusammenbau der Einzelelemente zum Gesamtsystem und c) die Lösung des resultierenden algebraischen Problems. Als Diskretisierungsgrundlage kommt meistens das Prinzip der stationären potentiellen Energie zum Einsatz (vgl. [1]); die dabei entstehende FEM-Variante ist also der wohlbekannten Deformationsmethode der Statik verwandt. Der aus praktischer Sicht schwerwiegende Nachteil dieser Methode besteht darin, daß Schnittkraftgrößen zwingenderweise mit einer geringeren Genauigkeit errechnet werden als die weitaus weniger interessierenden Verschiebungen. Einer der wesentlichen Aspekte dieser Arbeit ist darin zu sehen, daß durch die Formulierung eines verallgemeinerten Energieprinzips vom Typ Hellinger/Reissner eine Diskretisierungsgrundlage geschaffen wird, die es ermöglicht, Verschiebungs- und Schnittkraftgrößen mit der gleichen Genauigkeit zu berechnen.

2. Diskretisierung

Der Gesamtbereich des ebenen, beliebig gekrümmten Rahmentragwerkes wird in (n) Einzelelemente unterteilt. Zur knickfreien Annäherung der Tragwerksgeometrie im Ausgangs- und deformierten Zustand werden schwach gekrümmte Elemente (im Sinne der Marguerreschen Theorie) verwendet. Als Diskretisierungsgrundlage für die einzelnen Elemente dient ein modifiziertes, gemischt-hybrides, pseudokomplementäres Energiefunktional vom Typ Hellinger/Reissner. Zur Algebraisierung dieses Funktionals werden Ansatzfunktionen verwendet, welche folgende Nebenbedingungen a priori befriedigen müssen:

a) die Verschiebungsansatzfunktionen erfüllen die geometrischen Rand- und Übergangsbedingungen,
b) die Schnittkraftansatzfunktionen erfüllen den linearen Teil der inkrementellen Gleichgewichtsbedingungen des gekrümmten Balkens, d. h.:

$$\Delta^* N_{,x} = 0 ,$$

$$\Delta^* M_{,xx} + (\Delta^* N \cdot z^{(N)}_{,x})_{,x} = 0 .$$

Das hier vorgestellte Verfahren stellt somit formal eine Kombination der Deformationsmethode mit dem Kraftgrößenverfahren dar.

Mit Hilfe des sog. Fundamentallemmas der Variationsrechnung läßt sich zeigen, daß folgende Grundgleichungen vom Funktional a posteriori erfüllt werden (sog. Eulersche Gleichungen):

a) die geometrischen Verschiebungs-Verzerrungs-Relationen,
b) die vollständigen Gleichgewichtsbedingungen und
c) die statischen Rand- und Übergangsbedingungen.

Schnittkraftansatzfunktionen, die die statischen Übergangsbedingungen nicht a priori erfüllen, werden nach Pian als „hybrid" bezeichnet (vgl. [2]). Bekanntlich führt die Verwendung derartiger Ansätze dazu, daß ihre freien Parameter („Spannungskoeffizienten") auf Elementstufe zwischeneliminiert werden können, so daß der Zusammenbau der Einzelelemente zum Gesamtsystem ein algebraisches Gleichungssystem liefert, welches (wie dies ja auch bei der Deformationsmethode der Fall ist) nur Verschiebungen als Unbekannte enthält. Diese Tatsache ist nicht zuletzt deswegen erwähnenswert, weil:

— ein auf der Grundlage eines derartigen, gemischt-hybriden Energiefunktionals entwickeltes Element in ein nach der Deformationsmethode arbeitendes FEM-Programm eingebaut werden kann (vgl. Knothe [3]) und
— die überwiegende Mehrheit der auf dem Softwaremarkt erhältlichen Programmsysteme nach der Deformationsmethode arbeitet.

3. Behandlung des nichtlinearen Problems

Die einfachste Vorgehensweise zur Lösung von nichtlinearen Problemen ist wohl die „rein-inkrementelle" (in der numerischen Mathematik als Euler-Verfahren bekannt). Man geht dabei so vor, daß man das ursprünglich nichtlineare Problem linearisiert, d. h. man führt es auf eine Folge von linearen Problemen zurück. Evidenterweise hängt die Genauigkeit der dabei erzielten Lösung von der Anzahl der Linearisierungsschritte („Laststufen") ab.

Beim hier verwendeten Konzept zur Lösung des nichtlinearen Problems handelt es sich um ein „selbstkorrigierendes", inkrementelles Verfahren. Der Unterschied zur rein-inkrementellen Prozedur besteht darin, daß vor jeder Laststeigerung der (infolge der Linearisierung entstehende) numerische Fehler anhand der Eulerschen Gleichungen der verwendeten Variationsgleichung ermittelt und zu dem aufzubringenden Lastinkrement hinzuaddiert wird. Dieses Vorgehen stellt einen sehr effektiven Weg dar, um die Genauigkeit der erzielten Lösung zu verbessern, ohne dabei die Anzahl der Lastschritte erhöhen zu müssen. Eine stark vereinfachte Darstellung beider inkrementeller Verfahren ist der Abb. 1 zu entnehmen.

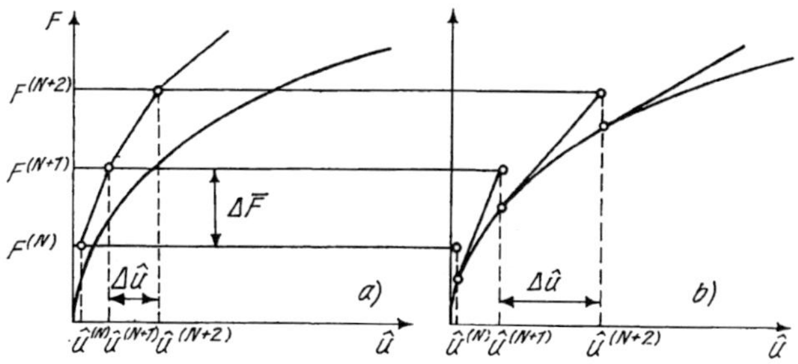

Abb. 1. Inkrementelle Lösungsverfahren
a) Euler-Verfahren, b) „selbstkorrigierendes" Verfahren

4. Numerische Ergebnisse

Zur Demonstration der Leistungsfähigkeit des entwickelten Finite-Element-Verfahrens werden nachfolgend die numerischen Ergebnisse für einige Testbeispiele vorgestellt:

a) Kragträger unter Gleichlast

Anhand dieses Beispieles wurde untersucht, ob die häufig verwendete Vorgehensweise zur Umrechnung von verteilten Lasten auf äquivalente Knotenlasten (d. h. Ansetzung des Prinzips der virtuellen Verrückungen am unverformten Tragwerk) zulässig ist oder nicht. Die Ergebnisse dieser Studie sind als Verformungsfiguren in Abb. 2 dargestellt, wobei die Bezeichnung „Gewichts"-Last bzw. Knotenlasten darauf hinweist, daß die Umrechnung am verformten bzw. unverformten Tragwerk durchgeführt wurde. Sie zeigen deutlich, daß die Art, wie die o. g. Lastenumrechnung vorgenommen wird, einen signifikanten Einfluß auf die erzielten numerischen Ergebnisse hat.

Als ein weiterer Effekt wurde schließlich der Einfluß von „Folgelasten" (hier: Druckbelastung) auf das Deformationsverhalten dieser einfachen Struktur studiert.

b) Kreisbogen unter Einzellast

Ziel der Untersuchung war hier, beide mögliche Knickformen des Tragwerks, die symmetrische (Durchschlagen) und die asymmetrische (Verzweigen) numerisch zu berechnen. Die erzielten Ergebnisse sind in Form von Last-Verschiebungs-Diagrammen in Abb. 3 dargestellt worden.

c) Asymmetrisch gelagerter Kreisbogenträger unter Einzellast

Anhand dieses in der Literatur bis dato kaum behandelten Beispiels sollte gezeigt werden, daß die Mitnahme von Korrektur-

[1]) Auszug aus der Dissertationsarbeit des Autors. Vortrag auf der GAMM-Tagung Berlin (West) 1980.

termen im Rahmen des inkrementellen Vorgehens (bei praktisch gleichbleibendem Rechenzeitaufwand) zu einer signifikanten Reduzierung des numerischen Fehlers führt (vgl. Abb. 4).

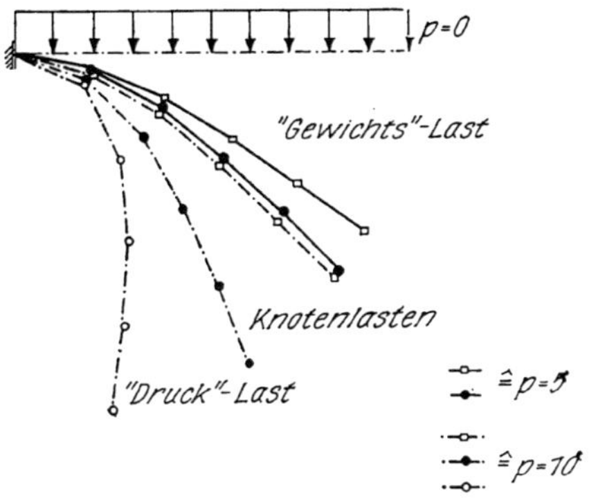

Abb. 2. Verformungsfiguren des Kragträgers unter Gleichlast

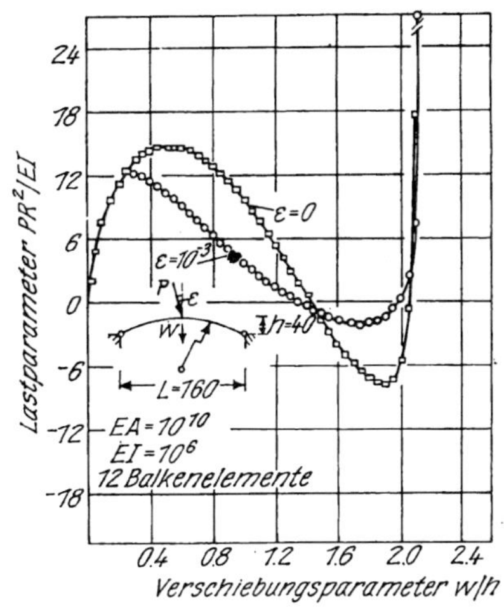

Abb. 3. Vollständiges Last-Verschiebungsdiagramm des Kreisbogens unter Einzellast

Literatur

1 ZIENKIEWICZ, O. C., The Finite Element Method, 3. Auflage, London 1977.
2 PIAN, T. H. H., A Historical Note about "Hybrid Elements", Int. J. Num. Meth. Eng. 13 (1978), S. 891–892.
3 KNOTHE, K., Ein Beitrag zur Systematisierung von Variationsprinzipien der Elastizitätstheorie und zu ihrer Verwendbarkeit für Finite-Element-Verfahren, ZAMM 58 (1978), S. 405–412.

Eingereicht am 12. 6. 1980

Anschrift: Dr.-Ing. DIMITRIOS KARAMANLIDIS, 1. Institut für Mechanik der TU Berlin, Sekr. MS 1, 1000 Berlin 10, Berlin (West)

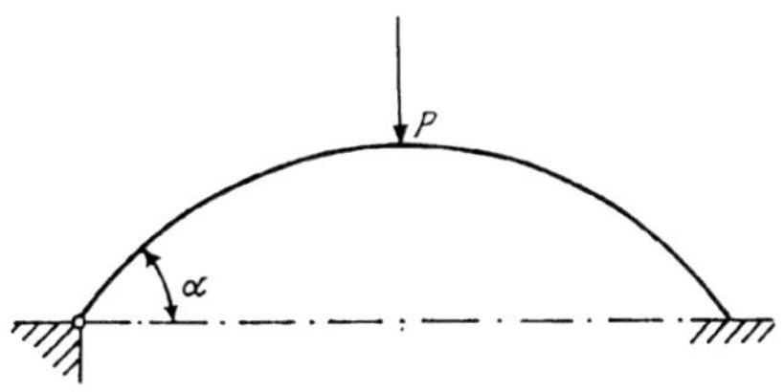

		α =	70°	80°	110°	120°
analytische Lösung		P_{krit}	1174,0	1078,0	758,0	447,0
FEM-Lösung	rein-inkrementell	P_{krit}	1125,60	983,88	512,0	332,74
		Anzahl der Lastschritte	51	55	40	58
		Fehler in %	4,12	8,73	32,45	25,56
	selbstkorrigierend	P_{krit}	1192,44	1088,95	751,44	451,20
		Anzahl der Lastschritte	63	77	85	47
		Fehler in %	1,57	1,02	0,87	0,94

Abb. 4. Kritische Lasten des asymmetrisch gelagerten Kreisbogens unter Einzellast

ZAMM 61, 204 – 205 (1981)

A. Granas / R. B. Guenther / J. W. Lee

A Note on the Thomas-Fermi Equation

1. Introduction

In 1927 L. H. Thomas and E. Fermi independently derived a boundary value problem for determining the electrical potential in an atom using the statistics for a degenerate gas. After making several changes of variables, their analysis led to the following *boundary value problem for an ionized atom*:
Find a function $y \geqq 0$ satisfying

$$x^{1/2}y'' = y^{3/2}, \qquad 0 < x < 1, \tag{1}$$

$$y(0) = a, \qquad y(1) = 0, \qquad a > 0 \quad \text{a constant}.$$

The treatment of this problem has been the subject of several investigations. See [5] and [6] and the references cited in their bibliographies. A modern derivation of (1) is given, e.g. in [1]. Physically, a must be one, although in order to present the numerical method, it is convenient to let $a > 0$ be arbitrary.

In this paper we show how the methods given in [3] can be used to treat not only problems similar to (1), but also more general singular boundary value problems. The analysis turns out to be of the utmost simplicity. In § 3, we also give theorems guaranteeing the convergence of the shooting method for numerically computing the solution. The notation is standard, although some of the terminology will be taken from [2] and [3].

2. The Thomas-Fermi problem and generalizations

Consider first the boundary value problem

$$x^{1/2}y'' = |y|^{1/2}\, y, \qquad y(0) = a, \qquad y(1) = 0, \tag{2}$$

and the related family of problems

$$x^{1/2}y'' = \lambda |y|^{1/2}\, y, \qquad y(0) = a, \qquad y(1) = 0. \tag*{(2)$_\lambda$}$$

Here $a > 0$ is a constant and λ is a parameter in $[0, 1]$. Because of the boundary condition, y'' must behave like $x^{-1/2}$ close to $x = 0$ so that we cannot expect to find a solution y which is twice continuously differentiable on $[0, 1]$.

Accordingly, denote by $\tilde{C}^2$ the space of functions, u, which are once continuously differentiable on $[0, 1]$, twice continuously differentiable on $(0, 1]$ and whose norm

$$||u|| = \max \left\{\max_{0 \leqq x \leqq 1} |u(x)|, \max_{0 \leqq x \leqq 1} |u'(x)|, \sup_{0 < x \leqq 1} |x^{1/2}u''(x)|\right\}$$

is finite.

Lemma 2.1: *$\tilde{C}^2$ equipped with the norm $||\cdot||$ is a Banach space.*

The proof of Lemma 2.1 is standard. We shall seek a solution y to (2) in the space $\tilde{C}^2$.

The set of functions in $\tilde{C}^2$ satisfying the boundary conditions in (2) is convex. By the existence theory presented in [3], to prove the existence of a solution to (2), it suffices to obtain a priori bounds on the solution to (2) in $\tilde{C}^2$ which are independent of $\lambda \in [0, 1]$ and to observe that when $\lambda = 0$, the unique solution to (2) is $y = a(1 - x)$.

The required a priori bounds are now contained in the following

Theorem 2.2: *Let y be a solution to* (2). *Then*
i) $0 \leqq y(x) \leqq a, \quad 0 \leqq x \leqq 1$;
ii) $|y'(x)| \leqq a + 2a^{3/2}/3$;
iii) $|x^{1/2}y''(x)| \leqq a^{3/2}$,
that is $||y|| \leqq 2a + 5a^{3/2}/3$.

A consequence of Theorem 2.2 is

Theorem 2.3: *There exists at least one solution to* (2) *and by* (i) *of* (2.2), *at least one solution to* (1).

Proof of 2.2: We may assume that $0 < \lambda \leqq 1$.

(i) Suppose there were values of x such that $y(x) < 0$. Let $x_0 \in (0, 1)$ be chosen so that $y(x_0) = \min y(x) < 0$. Then at x_0, the differential equation yields

$$0 \leqq y''(x_0) = x_0^{-1/2}\, |y(x_0)|^{1/2}\, y(x_0) < 0,$$

a contradiction. Similarly, we prove that $y(x) \leqq a$.

(iii) follows immediately from (i).

(ii) By the mean value theorem, there is a point $\tau \in (0, 1)$ such that $y'(\tau) = -a$. Consequently,

$$y'(x) + a = \int_\tau^x y''(t)\, dt = \int_\tau^x t^{-1/2}\, y''(t)\, t^{1/2}\, dt$$

whence (ii) follows from (iii).

Theorem 2.3 is an immediate consequence of 2.2. Furthermore, a standard, simple argument also yields the following uniqueness result.

Theorem 2.4: *There exists at most one solution in $\tilde{C}^2$ to the problem* (1).

These arguments can be generalized to obtain an existence and uniqueness theorem for the following problem.

We seek the solution to

$$y'' = \frac{\varphi(x, y)}{\psi(x)}, \qquad 0 < x < 1, \qquad y(0) = a, \qquad y(1) = b. \tag{3}$$

In the next theorem we shall make the following hypotheses on the functions φ and ψ:

(i) $\varphi\colon [0, 1] \times R \to R$ is continuous and there is a constant $M \geqq 0$ such that $y\varphi(x, y) > 0$ for $|y| > M$.

(ii) $\psi\colon [0, 1] \to [0, \infty)$ is continuous, $\psi > 0$ on $(0, 1)$, and $1/\psi$ is integrable on $[0, 1]$.

We denote by $\tilde{C}^2_\psi$ the Banach space of functions, u, which are once continuously differentiable on $[0, 1]$, twice on $(0, 1)$ and whose norm

$$||u|| = \max \{\max_{0 \leqq x \leqq 1} |u(x)|, \max_{0 \leqq x \leqq 1} |u'(x)|, \sup_{0 < x \leqq 1} |\psi(x)\, u''(x)|\}$$

is finite.

We now formulate the following theorem, the proof of which is nearly the same as the proof of Theorems 2.3 and 2.4.

Theorem 2.5: *There exists at least one solution to problem* (3). *If in addition $\partial\varphi/\partial y \geqq 0$ on $(0, 1) \times R$, then this solution is unique.*

Remark 1: After 2.5 has been proven, the assumption (i) can be relaxed to $y\varphi(x, y) \geqq 0$ for $|y| > M$ by a standard limiting argument.

Remark 2: If in (3), φ depends upon (t, y, y'), then existence in $\tilde{C}^2_\psi$ can be proven provided that the assumption (i) is replaced by

(i′) There is a constant $M \geqq 0$ such that $y(\varphi x, y, 0) > 0$ for $|y| > M$. Furthermore, $\varphi(t, y, p)$ is a bounded function on $[0, 1] \times [-M, M] \times R$.

3. Shooting for the Thomas-Fermi equation

In this section we restrict ourselves to the Thomas-Fermi problem (1), although the discussion does carry over to the problem (3). The theoretical description and discussion of the shooting method parallels that given in [4]. The idea is to consider the Cauchy problem

$$u'' = x^{-1/2}\, |u|^{1/2}\, u, \qquad u(1) = 0, \qquad u'(1) = \alpha \tag{4}$$

and choose α so that $u(x, \alpha)$ satisfies (2) and hence (1). Such an α always exists since (1) has a unique solution. It can be computed as follows. Let $\beta > 0$ be arbitrary. Then (1) can be solved with $a = \beta$ to obtain a function $y(x, \beta)$. Set $y'(0, \beta) \equiv g(\beta) = \alpha$. A repeat of the argument in [4] yields

Theorem 3.1:

(i) *g is a strictly increasing, differentiable function defined on $(0, \infty)$ whose range is an open interval J. Let $h\colon J \to (0, \infty)$ be the inverse of g. Then $\beta = h(\alpha)$ is strictly increasing, continuously differentiable, and $h'(\alpha) > 0$ for α in J.*

(ii) *For all α in J, the Cauchy problem* (4) *has a unique solution $u(x, \alpha)$ defined for all x in $[0, 1]$; namely*

$$u(x, \alpha) = y(x, h(\alpha)).$$

(iii) *For each $a > 0$, there is an α in J such that $u(0, \alpha) = a$.*

In order to calculate α, we solve $u(0, \alpha) = a$ by Newton's method. Let $\alpha_0 > 0$ be an initial guess. Compute

$$\alpha_{n+1} = \alpha_n - [u(0, \alpha_n) - a]/u_\alpha(0, \alpha_n), \qquad n = 0, 1, \ldots, \tag{5}$$

where $u_\alpha = \partial u/\partial\alpha$. First of all $u_\alpha(0, \alpha) \neq 0$. For $u_\alpha(0, \alpha)$ satisfies

$$u''_\alpha = (3/2)\, u^{1/2} u_\alpha, \qquad u_\alpha(1) = 0, \qquad u'_\alpha(1) = 1,$$

and if $u_\alpha(0) = 0$, $u_\alpha \equiv 0$ contradicting the fact that $u'_\alpha(1) = 1$.

We can, therefore, formulate the following theorem, the proof of which is completed exactly as in [4].

Theorem 3.2: *Let y be a solution to* (1). *Then there is a $\delta > 0$ such that if $|\alpha_0 - y'(0)| < \delta$, then for each of the Newton* *iterates $\alpha_1, \alpha_2, \ldots$ defined by* (5) *the Cauchy problem* (4) *has a unique solution $u(x, \alpha_n)$ which converges to $y(x)$ in $\tilde{C}^2$.*

From a theoretical standpoint, Theorems 3.1 and 3.2 tell us that we can always shoot successfully. From a computational standpoint, however, it would appear that some care must be exercised in implementing 3.1 and 3.2 since if the grid is very small, the computations would perhaps be overwhelmed when one approaches zero. The following graph was constructed by taking 101 grid points, shooting to .01 using the RUNGE-KUTTA method and the extrapolating to $x = 0$ using a standard third order method. The numbers agree with those given in [6]. The justification of such a simple procedure, however, is rather tedious and will be treated in a subsequent paper.

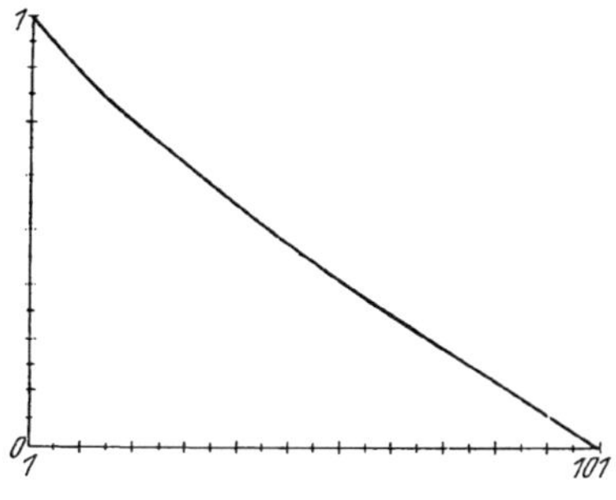

Fig. 1

Bibliography

1 EISBERG, R. M., Fundamentals of Modern Physics, John Wiley & Sons, New York-London-Sydney 1961.
2 GRANAS, A., Sur la méthode de continuité de POINCARÉ, C.R. Acad. Sci., Paris **282** (1976), 983–985.
3 GRANAS, A.; GUENTHER, R. B.; LEE, J. W., On a theorem of S. BERNSTEIN, Pacific J. Math. **74** (1978), 67–82.
4 GRANAS, A.; GUENTHER, R. B.; LEE, J. W., The shooting method for the numerical solution of a class of nonlinear boundary value problems, SIAM J. Numer. Analysis, to appear in 1979.
5 DE HOOG, F. R.; WEISS, R., The numerical solution of boundary value problems with an essential singularity, SIAM J. Numer. Analysis **16** (1979), 637–669.
6 MOONEY, J. W., Monotone methods for the THOMAS-FERMI equation, Quart. Appl. Math. **36** (1978), 305–314.

Eingereicht: 19. 10. 1979.

Anschriften: Professor Dr. ANDRZEJ GRANAS, Department of Mathematics, University of Montreal, Montreal, Que. H3C 3J7, Canada;
Professor Dr. RONALD B. GUENTHER, Professor Dr. JOHN W. LEE, Department of Mathematics, Oregon State University, Corvallis, OR. 97331, USA

BUCHBESPRECHUNGEN

Löthe, H. / Müller, K. P., Taschenrechner. Erläuterungen zum Aufbau und Vorschläge für den Unterricht. Stuttgart, B. G. Teubner 1979. 168 S., 49 Abb., DM 18,80. ISBN 3-519-02711-9

Zum Thema Taschenrechner in der Schule bietet das vorliegende Büchlein Beispiele, die bestimmte Funktionsweisen motivieren und erläutern sollen, systematische Gesichtspunkte und Beispiele für den Unterricht. Als durchaus erwünschte Nebenwirkung des Taschenrechnereinsatzes in der Schule erwarten die Verfasser, daß der Physikunterricht besser auf Inhalte des Mathematikunterrichtes zurückgreifen kann, als dies bei einer nur strukturorientierten Mathematik möglich ist. Die Frage, welche Form mathematischen Denkens in Zukunft dem Schulunterricht zu Grunde gelegt werden soll: funktionales Denken, Denken in Strukturen oder algorithmisches Denken, sollte in der Tat, wie es die Autoren zu bedenken geben, so entschieden werden, daß die Anwendung die geeignete Form der Mathematisierung bestimmt.

Berlin J. FÖRSTE

Karl, H. (Hrsg.), Datennetze, 1. Treffen 1978 des German Chapter of the ACM am 2. 6. 1978 in Wiesbaden. Applied Computer Science — Berichte zur praktischen Informatik 15. München-Wien, Carl Hanser Verlag 1979. 200 S., 70 Abb., DM 38,—. ISBN 3-446-13016-0

Das Gebiet der Rechnernetze entwickelte sich in den letzten Jahren zu einer Schwerpunktforschungslinie der Informatik. Das vorliegende Buch bringt eine Reihe von Vorträgen eines Treffens des German Chapters of the ACM zu dieser Problematik.

Der Beitrag von WITT demonstriert die Realisierung einer Systemtransparenz durch Überschichtung von Benutzerschnittstellen bei heterogenen Systemen. Das Rechnerverbundkonzept der Firma Hewlett Packard stellt MUND vor. Aspekte verteilter Datenbanken in einem Informationssystem werden von KARL behandelt. Einen Überblick über Zielstellungen, Kommunikationsfunktionen und Netzkonfigurationen vermittelt STROBL. Ein implementiertes Modell zur Simulation der Netzbelastung beim Einsatz verteilter Datenbanken wird von BIEBER u. FLOREK beschrieben. BUTSCHER u. HEINZE berichten über die Gestaltung und Implementierung eines File-Transfer-Protokolls für das HMI-Netzwerk. Effizienz- und Overheaduntersuchungen von Datenzugriffssystemen für dieses Netzwerk werden von POPESCU-ZELETIN u. BUTCHER durchgeführt. Entwurfsgrundlagen und Dienste des Berlin(West)-Rechnernetzwerks BERNET werden von HEINZE u. a. erläutert. Die stufenweise Implementierung eines Kommunikationssystems für die Universität Erlangen-Nürnberg umreißt REINARTZ. Das Konzept einer virtuellen Netzwerkmaschine zur Programmierung eines Rechnernetzwerks für spezielle Anwendungen wird von KEMEN untersucht. HECKL weist auf einige Fragen des interaktiven Entwurfs von Nachrichtennetzen hin. Rechnergestützter Entwurf und Fertigung stellen verschiedene Anwenderanforderungen an den Rechnerverbund, die von GRAUER zusammengestellt werden. Zur Überwindung von Schwächen konventioneller Rechnerstrukturen schlägt LÜKEN eine Multiprozessor-Struktur mit einem autonomen Datenübertragungssystem vor.

Mit den vorliegenden Beiträgen vermittelt das Buch einen Einblick in den Stand der Arbeiten zu einigen Netzwerkprojekten.

Zeuthen W. MEIER

Herschel, R. / Pieper, F., PASCAL, Systematische Darstellung von PASCAL und CONCURRENT PASCAL für den Anwender. Reihe Datenverarbeitung. München-Wien, R. Oldenbourg. 1979. 248 S., 183 Abb., DM 23,80. ISBN 3-486-23681-4

Die Sprache PASCAL ist eine der modernsten höheren Programmierungssprachen, konzipiert von NIKOLAUS WIRTH, und berücksichtigt besonders die Forderungen der strukturierten Programmierung. Die einzelnen Sprachelemente von PASCAL werden im Buch in einer sehr ansprechenden und gut verständlichen Art erklärt, und insbesondere wird ihre Anwendung bei der Formulierung von Algorithmen gezeigt. Die gewählten Beispiele, mehrere sind nichttrivial, lassen die Vorteile der Benutzung von PASCAL deutlich erkennen. Bei der Darlegung der Sprachelemente nehmen die Datentypen, entsprechend ihrer Bedeutung, einen breiten Raum ein. Ausführlich besprochen werden der Zeigertyp und bei den strukturierten Datentypen die Felder, Records, Files und auch der Mengentyp. — Im zweiten Teil des Buches wird die Sprache CONCURRENT PASCAL, sie geht auf BRINCH HANSEN zurück, vorgestellt. Mit CONCURRENT PASCAL ist es möglich, parallele Prozesse, die immer häufiger behandelt werden, zu beschreiben. Zunächst werden die hinzukommenden Sprachelemente und die Grundlagen und Verfahren der Programmierung paralleler Prozesse, wie die Systemtypen Prozesse, Monitore und Klassen, die Unterprogramme und die Zugriffsrechte besprochen und an verschiedenen Beispielen, sie sind entsprechend der schwierigeren Natur der parallelen Prozesse nicht einfach darlegbar, erläutert. Abschließend wird auf die Behandlung verschiedener Probleme, wie das Koordinierungsproblem, die autonome Ein- und Ausgabe, die Werteschlangenproblematik und das Feldrechner-Modell, mit CONCURRENT PASCAL eingegangen. — Die gemeinsame Darstellung von PASCAL und CONCURRENT PASCAL in einem Buch ist sehr gut gelungen, wobei besonders die Angabe vieler gut gewählter Beispiele hervorgehoben werden soll.

Berlin F. GRUND

Mardia, K. V. / Kent, J. T. / Bibby, J. M., Multivariate Analysis. London-New York-Toronto-Sydney-San Francisco, Academic Press 1979. XV, 521 S., $ 34.00 P/B. ISBN 0-12-471252-5

This excellently written monograph deals with modern developments of multivariate analysis in statistics whereas the data analytic as well as the theoretical aspect is contained. Supported by many practical and numerical examples the book is readable for practitioners, too. After some preparing chapters on generalizations of univariate distributions and test procedures the wide-spread field of multivariate investigations in linear statistics is handled. Chapters 6 and 7 are devoted to the multivariate regression model and the econometric model. Chapters 8 to 12 handle the principal component analysis, factor analysis, canonical correlation, discrimination and the analysis of variance. Cluster analysis, the multi-dimensional scaling and so-called directional data are given in a way which is very useful for problems in practice. The appendix on matrix results is helpful for proofs within the book.

Berlin H. TOUTENBURG

Cottle, R. W. / Giannessi, F. / Lions, J.-L., Variational Inequalities and Complementarity Problems. Theory and Applications. Chichester-New York-Brisbane-Toronto, John Wiley & Sons 1980. XVII, 408 S., £ 17.75. ISBN 0-471-27610-3

Das Buch enthält die Proceedings einer im Juni 1978 in Erice (Sizilien) durchgeführten Internationalen Schule über Variationsungleichungen und Komplementaritätsprobleme in der mathematischen Physik und der Ökonomie. Es ist dem Andenken an den im April 1978 verstorbenen Prof. G. STAMPACCHIA gewidmet.

Das erklärte Ziel der Veranstaltung war es, den Austausch zwischen Forschern auf dem Gebiet der Variationsungleichungen und solchen auf dem Gebiet der Komplementaritätsprobleme zu fördern. Diese Gebiete haben sich trotz ihres inhaltlichen Zusammenhangs bisher weitgehend unabhängig voneinander entwickelt. Dementsprechend gehen nur wenige der Arbeiten des Bandes auf den Zusammenhang zwischen Variationsungleichungen und Komplementaritätsproblemen ein. Variationsungleichungen sind Aufgaben folgender Form:

(1) Gesucht wird $u \in K$ derart, daß $\langle Au, v - u\rangle = 0$ für alle $v \in K$. Dabei bezeichnet K eine abgeschlossene konvexe Menge im Raum V (i. a. ein HILBERT-Raum) und A eine Abbildung von V in den dualen Raum V^*. Komplementaritätsprobleme können wie folgt formuliert werden:

(2) Gesucht wird $u \in K$ derart, daß $Au \in K^*$ und $\langle Au, u\rangle = 0$. Hier ist K ein konvexer abgeschlossener Kegel in V und $K^* := \{f \in V^* \mid \langle f, v\rangle = 0 \text{ für } v \in K\}$ der polare Kegel in V^* (A und V wie in (1)).

Ist K ein Kegel, so ist u genau dann Lösung von (2), wenn es Lösung von (1) ist. Das bedeutet, daß man Komplementaritätsprobleme als spezielle Variationsungleichungen auffassen kann. Umgekehrt lassen sich viele Variationsungleichungen in Komplementaritätsprobleme transformieren.

Dem Anliegen der Veranstaltung entsprechend sind die meisten Arbeiten Übersichtsartikel bzw. Referate über Ergebnisse, die an anderer Stelle ausführlich dargelegt worden sind. Die Übersicht über den Inhalt des Buches wird durch ein detailliertes Inhaltsverzeichnis und ein Sachwortverzeichnis erleichtert. Der Band enthält folgende Arbeiten: J. L. LIONS, The Work of G. STAMPACCHIA in Variational Inequalities; C. BAIOCCHI, Variational Inequalities and Free-Boundary Problems; H. BERESTYCKI, P. L. LIONS, Existence of a Ground State in Non-Linear Equations of the KLEIN-GORDON Type; H. BRÉZIS, Some Variational Problems of the Thomas-Fermi Type; I. CAPUZZO DOLCETTA, U. MOSCO, Implicit Complementarity Problems and Quasi-Variational Inequalities; R. CONTI, A Minimum Time Problem; R. W. COTTLE, Some Recent Developments in Complementarity Theory; M. CUGIANI, Some Stochastic Strategies in Optimization Problems; G. B. DANTZIG, P. L. JACKSON, Pricing Underemployed Capacity in a Linear Economic Model; A. FRIEDMAN, The Dam Problem with Variable Permeability; A. FRIEDMAN, Stochastic Control with Partial Observations; F. GIANESSI, Theorems of Alternative, Quadratic Programs, and Complementarity Problems; F. J. GOULD, C. P. SCHMIDT, An Existence Result for the Global NEWTON Method; D. KINDERLEHRER, The Smoothness of Free Boundaries Governed by Elliptic Equations and Systems; C. E. LEMKE, A Survey of Complementarity Theory; P. L. LIONS, An Estimate of the LIPSCHITZ-Norm of Solutions of Variational Inequalities and Applications; L. McLINDEN, The Complementarity Problem for Maximal Monotone Multifunctions; U. MOSCO, On Some Non-Linear Quasi-Variational Inequalities and Implicit Complementarity Problems in Stochastic Control Theory; M. RAOUS, On Two Variational Inequalities Arising from a Periodic Viscoelastic Unilateral Problem; R. T. ROCKAFELLAR, LAGRANGE Multipliers and Variational Inequalities; F. SCARPINI, A Numerical Approach to the Solution of Some Variational Inequalities; F. SCARPINI, Error Estimates for the Finite Element Approximation of Some Variational Inequalities; J. SIJBRAND, Bifurcation Analysis of the Free Boundary of a Confined Plasma; R. TEMAM, Mathematical Problems in Plasticity Theory; R. J. B. WETS, Convergence of Convex Functions, Variational Inequalities and Convex Optimization Problems.

Berlin K. GRÖGER

Dynkin, E. B. / Yushkevich, A. A., Controlled Markov Processes. Grundlehren der mathematischen Wissenschaften 235. Berlin-Heidelberg-New York, Springer-Verlag 1979. XVII, 289 S., 36 Abb., DM 79,50. US $ 46.90. ISBN 3-540-90387-9

Controlled MARKOV processes arise in many diverse fields. This book is a systematic presentation of the theory of MARKOV control processes with discrete time. Applications of this theory are discussed to various concrete problems. Particular attention is paid to control models of economic planning, taking account of stochastic factors. A reader interested in the applications can get through the book with a knowledge of elementary probability theory. To understand all parts of the book it is necessary to have a background in measure theory, metric spaces and functional analysis.

An advantage of the presentation of this book is that it starts from the simplest case where the theory already begins to work and goes step by step to the most general situation. As a rule, proofs are first given for special cases and followed by generalizations.

The main chapters of the book are: Control on a finite time interval: Finite and Denumerable Models, Semicontinuous Models, General (BOREL) Models; Control on an Infinite Time Interval: Discrete Models, BOREL Models, Homogeneous Models, Maximization of the Average Reward per unit Time; Some Applications: Models with incomplete information. Concave

Models, Models of Economic Development. The appendix contains results about BOREL spaces, analytic sets, theorems on measurable selection, conditional distributions and some Lemmas on measurability. A detailed list of references is given at the end of the book.

Berlin E. PLATEN

Whitham, G. B., Lectures on Wave Propagation. Berlin-Heidelberg-New York, Springer-Verlag 1979. Bombay, Tata Institute of Fundamental Research 1979. VII, 148 S., 31 Abb., DM 18,—. US $ 8.00. ISBN 3-540-08945-4

With these notes of a two-month course on wave propagation the author gives an introduction to methods and techniques of applied mathematics. After some background from his book "Linear and nonlinear waves", Wiley Interscience, New York (1974) the material is especially dedicated to water waves on a sloping beach. For this special field the construction of solutions of the wave equations is demonstrated for equations with growing complexity. Finally for certain fundamental equations like KORTEWEG-DE VRIES equation exact solutions are derived by means of perturbation methods.

In the referee's opinion this is a well-reading and stimulating contribution to the literature on wave equations.

Berlin G. LIPPOLD

Pedley, T. J., The Fluid Mechanics of Large Blood Vessels. Cambridge Monographs on Mechanics and Applied Mathematics. Cambridge-New York-Melbourne, Cambridge University Press 1980. XV, 446 S., ISBN 0-521-22626-0

Das Interesse von Mechanikern und Mathematikern an den Problemen der Blutzirkulation hat schon früh bestanden, lange bevor die Kreislaufforschung durch die Volkskrankheit „Infarkt" in den Blickpunkt der Öffentlichkeit geraten ist. Daß dabei eine Reihe von wesentlichen Erkenntnissen der Mechanik durch das Streben nach einem Verständnis für die physikalischen Vorgänge in den Blutgefäßen gewonnen wurde, ist evident. Da sich ein ständig wachsender Kreis von Mechanikern und Mathematikern den Fragen der Strömungsmechanik des Herz-Kreislaufsystems widmet, ist die Publikation einer, wie sie der Verfasser nennt, Forschungsmonographie zu diesem Themenkreis mit dem Schwerpunkt bei der mathematischen Behandlung des Gegenstandes nur zu begrüßen. Die Beschränkung auf die großen Gefäße (Herz, Arterien, Venen) erfolgt deshalb, weil bei ihrer Behandlung von der Mikrostruktur des Blutes abgesehen werden kann und ein NEWTONsches Verhalten (wenigstens über den größten Teil der Zeit eines Schlages) postuliert werden darf; als weitere Gründe werden die in der Mikrozirkulation wesentlichen aktiven Änderungen der Arteriolendurchmesser und das Fehlen von experimentellem Vergleichsmaterial angeführt. Doch auch so bleibt der Gegenstand schwierig genug, denn die Erscheinungen der Instationarität, der Flexibilität der Rohrwand und der nichtlinearen Bettung der Gefäße bleiben erhalten. Die Darstellung ist so geordnet, daß in einer Einleitung die physiologischen Eigenheiten des Herz-Kreislaufsystems erläutert werden, wobei bereits auf die strömungsmechanischen Probleme hingewiesen wird. Daran schließt sich eine Darstellung der Druckimpulsfortpflanzung in ein- und zweidimensionaler Behandlung u. a. unter Berücksichtigung von Verengungen und Verzweigungen und einer Bettung mit viskoelastischer Dämpfung an. In den Kapiteln 3 bis 5, die das Kernstück des Buches bilden, wird auf die Berechnung der Geschwindigkeitsprofile und der Wandschubspannungen in Arterien eingegangen. Im Kapitel 3 werden die Fälle betrachtet, die sich auf die Strömung in geraden Rohren zurückführen lassen. Dabei werden Lösungen für die stationäre Strömung und die instationäre Strömung ohne und mit Rückströmung diskutiert, deren Grundlage die Voraussetzung einer gegenüber dem Krümmungsradius der Rohrwand verschwindenden Grenzschichtdicke ist, so daß die Behandlung der Strömungsprobleme mit Hilfe der Grenzschichtgleichungen für ebene Strömungen erfolgen kann. Im Kapitel 4 werden zur Berechnung der Strömungsverhältnisse in gekrümmten Arterien Lösungen der Bewegungsgleichungen für die instationäre zähe Strömung im gekrümmten Rohr unter der Voraussetzung des vollausgebildeten Zustandes und einer kleinen Krümmung der Rohrachse angeboten, wobei der oszillatorische Fall und die Anfahrbewegung behandelt werden. Die Krümmung der Rohrachse darf dabei auch leicht variieren. Das am wenigsten theoretisch gesicherte Gebiet ist das in Kapitel 5 behandelte Gebiet der Verzweigungen. Hier liegen bisher nur wenige Lösungen für stationäre Strömungen vor, so daß das Modellexperiment noch am ehesten eine Vorstellung von den Vorgängen liefert. Sonderfälle, wie der Abfluß in einen extrem kleinen Seitenarm lassen sich exakt mit den Stokesschen Gleichungen behandeln. Das Problem der Instabilität und der Turbulenz tritt allein für den Bereich der Aorta auf. Die bisherigen Untersuchungen verweisen allerdings auf eine Instabilität während der Spitze der Systole. Die möglichen Mechanismen für die Entstehung der Instabilität werden diskutiert. Den Inhalt des letzten Kapitels bilden die Strömungen durch kollabierende Venen. In einem Anhang wird eine Analyse des bei den meisten Messungen benutzten Heißfilmanemometers gegeben. Das mit Akribie und profunder Sachkenntnis geschriebene Buch dürfte ein Standardwerk der Bioströmungsmechanik werden. Darüber hinaus vermag es aber selbst denen, die dem Problemkreis ferner stehen, eine Fülle von Anregungen zu geben, sich dieser oder jener der vielen noch offenen interessanten Fragen zu widmen.

Berlin K. REMKE

Pacault, A. / Vidal, C. (Eds.), Synergetics far from Equilibrium. Proceedings, Bordeaux, France, 1978. Springer Series in Synergetics 3. Berlin-Heidelberg-New York, Springer-Verlag 1979. IX, 175 S., 109 Abb., DM 59,—. US $ 32.50. ISBN 3-540-09304-4

Die Synergetik untersucht kooperative Effekte auf einem phänomenologischen Niveau, so daß sich makroskopisch ähnliche Verhaltensweisen ergeben bei komplexen Systemen ganz unterschiedlicher mikroskopischer Struktur. Diese Beobachtung gibt Anlaß zur Erarbeitung einheitlicher begrifflicher und mathematischer Methoden für eine gemeinsame, vergleichende Analyse solcher Erscheinungen. Im vorliegenden dritten Band der dieser Problematik gewidmeten „Springer Series in Synergetics" werden an chemischen, biochemischen und physikalischen Systemen Reaktionen und Eigenschaften untersucht, welche sich aus der Existenz mehrfacher stationärer Zustände sowie mehr oder weniger umfangreicher Hierarchien von Ordnungsparametern und Instabilitäten ergeben. Hierdurch erfaßbar werden Phänomene wie nichtlineare Oszillationen, dissipative Strukturen, Nichtgleichgewichtsphasenübergänge, Übergänge von regulären in erratische Zustände (Chaos, Turbulenz).

Die mathematische Beschreibung dieser Reaktions-Diffusionssysteme erfolgt durch Systeme nichtlinearer gewöhnlicher bzw. partieller Differentialgleichungen, bei denen Lösungsverzweigungen (Bifurkationen) und „seltsame (hyperbolische) Attraktoren" existieren. Von besonderem mathematischen Interesse sind die im Beitrag von RÖSSLER behandelten Zusammenhänge zwischen seltsamen Attraktoren und Diffeomorphismen (bei verschiedenen Dimensionszahlen ($n = 3$, $n = 4$)).

Bei den physikalischen Systemen spielt der Laser eine wichtige Rolle wegen seiner Übersichtlichkeit trotz seines großen Spektrums an Phänomenen, die typisch sind unter dem Gesichtspunkt der Synergetik (vgl. Beitrag HAKEN). Bei den hydrodynamischen Systemen werden u. a. untersucht: durch den SORET-Effekt getriebene Instabilitäten (VELARDE) sowie der Übergang von der RAYLEIGH-BÉNARD-Instabilität zur Turbulenz (DUBOIS und BERGÉ).

Von allgemeinem mathematischen wie praktischen Interesse sind schließlich die Beiträge über den Einfluß von Fluktuation (Rauschen) auf kooperative Effekte (LEFEVER u. HORSTHEMKE, DE KAPPER und HORSTHEMKE, SUZUKI).

Der vorliegende Band stellt eine erfolgreiche Fortsetzung der genannten Reihe sowie der auf H. HAKEN zurückgehenden Konzeption dar (vgl. auch Naturw. *67*, 121 (1980)).

Berlin G. WALLIS

Schumann, W. / Dubas, M., Holographic Interferometry. From the Scope of Deformation Analysis of Opaque Bodies. Springer Series in Optical Science 16. Berlin-Heidelberg-New York, Springer-Verlag 1979. X, 194 S., 73 Abb., DM 59,—. US $ 32.50. ISBN 3-540-09371-0

Der vorliegende Band 16 der Reihe über optische Wissenschaften schlägt eine Brücke zwischen Optik und Mechanik. Die Verfasser haben sich das Ziel gesetzt, die holografische Interferometrie von der Warte der Anwendung zur Deformationsanalyse undurchsichtiger Körper zu behandeln. Dazu bedienen sie sich in konsequenter Weise der Darstellung mit Hilfe der Differentialgeometrie und der Tensoranalysis. Nach einer kurzen Einführung in diese Gebiete werden die Prinzipien der Bildentstehung in der Holografie behandelt. Dabei gehen die Autoren besonders auf den Einfluß von Veränderungen während der holografischen Rekonstruktion auf die Lage des rekonstruierten Bildes ein. Auf Grund dieser Darstellung wird die Interpretation der Interferenzstreifen der holografischen Interferogramme, die den dritten wesentlichen Teil des Buches bildet, ausgeführt. Es wird von Beginn an die Tatsache berücksichtigt, daß die Ermittlung von Verschiebungen an den Interferogrammen für den Festkörpermechaniker meistenteils nur ein Zwischenschritt ist, da letztlich Dehnungen oder Krümmungen gefragt sind. Hierbei werden auch die Eigenschaften der Interferenzstreifen, ihr Abstand, ihre Richtung und ihre Lokalisierung, die mit den ersten Ableitungen der optischen Wegdifferenzen im Zusammenhang stehen, betrachtet und gezeigt, wie deren Dehnungen und Verdrehungen bestimmt werden können. Mit der Behandlung dieser Probleme haben die Autoren eines der aktuellsten Gebiete der Anwendung der holografischen Interferometrie erstmalig in geschlossener Form dargestellt. In einem abschließenden Kapitel wird der Zusammenhang zwischen den zweiten Ableitungen der Verschiebungen und der optischen Weglängen-Differenzen betrachtet und durch Hinweise auf mögliche Anwendungen ein Weg für weitere Entwicklungen gezeigt. Durch Verweis auf ca. 370 Literaturstellen wird dem Leser der Zugang zu weiterer Originalliteratur ermöglicht. Das Buch hat seinen besonderen Vorzug in der durchgehenden Geschlossenheit der Darstellung. Es ist dadurch zwar keine Einführung in die experimentellen Grundlagen der holografischen Interferometrie, wird aber für den auf diesem Gebiet bereits tätigen, erfahreneren Physiker und Ingenieur eine wichtige Ergänzung sein.

Berlin G. Wernicke

In der DDR beziehbar im Rahmen des Kontingents.

von Bally, G. (Ed.), Holography in Medicine and Biology. Proceedings of the International Workshop, Münster, 1979. Springer Series in Optical Sciences 18. Berlin-Heidelberg-New York, Springer-Verlag 1979. IX, 269 S., 224 Abb., DM 78,–. US $ 42.90. ISBN 3-540-09793-7

Das Buch stellt in thematischer Gliederung die Beiträge eines internationalen Symposiums vor, das 1979 in Münster stattfand. Die wesentlichen Kapitel sind dabei die Anwendung der Holografie in der Orthopädie, der Ophthalnologie, der Urologie, der Hals-Nasen-Ohren- und der Zahnmedizin, Anwendungen in der Radiologie und der Biologie, sowie Fragen der Moiré-Topografie, der akustischen Holografie und spezielle holografische Techniken. Obwohl vom Titel her eher eine starke Spezialisierung auf Medizin und Biologie erwartet wird, bietet der Band doch interessante und wichtige Beiträge auch für den experimentell tätigen Mechaniker und Biomechaniker, da sich ein großer Teil der Vorträge mit Fragen der dreidimensionalen Abbildung, der Vermessung von gekrümmten Oberflächen und der Verformungsanalyse mit Hilfe der holografischen Interferometrie beschäftigt. Auch hier zeigt sich, daß die Verbindung der klassischen Interferometrie mit den holografischen Aufzeichnungsverfahren der experimentellen Forschung wesentliche Impulse zu geben vermag. Jedes einzelne Gebiet wird durch einen Übersichtsartikel eingeleitet, dem einige Originalbeiträge folgen. Beim aufmerksamen Lesen der einzelnen Beiträge ist für die experimentelle Arbeit eine große Zahl von Hinweisen zu finden, die in dieser Form sonst nur aus einzelnen Beiträgen in verschiedenen Zeitschriften zu finden sind. In dieser Hinsicht ist die Herausgabe dieses Bandes von nicht zu unterschätzender Bedeutung. Die große Zahl informativer Abbildungen tragen zum guten Verständnis bei, Literaturhinweise zu den einzelnen Beiträgen ermöglichen noch tieferes Eindringen in die Teilgebiete.

Berlin G. Wernicke

FSC
www.fsc.org
MIX
Papier | Fördert gute Waldnutzung
FSC® C083411

Zeitfracht Medien GmbH
Ferdinand-Jühlke-Straße 7
99095 Erfurt, Deutschland
produktsicherheit@kolibri360.de